Shift

ALSO BY ETHAN KROSS

Chatter

Shift

*Managing Your Emotions—
So They Don't Manage You*

ETHAN KROSS

CROWN
NEW YORK

CROWN
An imprint of the Crown Publishing Group
A division of Penguin Random House LLC
crownpublishing.com

Library of Congress Cataloging-in-Publication Data
Names: Kross, Ethan, author.
Title: Shift : managing your emotions - so they don't manage you / by Ethan Kross.
Description: New York : Crown Publishing, [2025] |
Includes bibliographical references and index.
Identifiers: LCCN 2024039431 (print) | LCCN 2024039432 (ebook) | ISBN
9780593444412 (hardcover) | ISBN 9780593444429 (ebook)
Subjects: LCSH: Emotions—Psychological aspects.
Classification: LCC BF531 .K76 2025 (print) | LCC BF531 (ebook) | DDC
152.4—dc23/eng/20241121
LC record available at https://lccn.loc.gov/2024039431
LC ebook record available at https://lccn.loc.gov/2024039432

Hardcover ISBN 978-0-593-44441-2
International edition ISBN 979-82-17-08621-4
Ebook ISBN 978-0-593-44442-9

Editor: Gillian Blake
Editorial assistants: Amy Li and Jess Scott
Production editor: Ashley Pierce
Text designer: Amani Shakrah
Production manager: Heather Williamson
Proofreaders: Sarah Rutledge and Kevin Clift
Indexer: Ken DellaPenta
Publicist: Mary Moates
Marketer: Mason Eng

Manufactured in the United States of America

1 2 3 4 5 6 7 8 9

First Edition

To Bubby, for (inadvertently) guiding me to "ask why,"
Mom, for always being there,
and Lara, Maya, and Dani . . . for everything else.

Contents

Why Is a Crooked Letter

As the farmhouse door flew open and the soldiers stormed in, Dora realized her first warm meal in months was actually a trap.

It was winter 1943. Dora Kremin and her boyfriend, Izzy, had been living in the snowy woods outside Eishyshok, in what was then Poland. They were a young couple in their twenties. Before the war their lives had been all about family, community, Friday night meals with music and games: a normal life in a little town. Then the Nazis came, on motorcycles. They told the Jewish townspeople that everything would be normal—that they would all work together, live together. But soon enough, it started: Yellow stars. Forced labor. Houses looted. People rounded up like animals.

One day, in the market square where she used to buy eggs and cheese as a child, Dora was lined up to be shot along with her family, friends, and neighbors she'd known her whole life.

Somehow, Izzy convinced one of the Polish guards he recognized from the days before the occupation to let her run away.

And she ran.

Her sister and brother-in-law also made it out of Eishyshok, and the four of them, along with a small group of other Jews, found refuge in a bunker in the forest with a band of partisan resistance fighters, after brief stops at multiple ghettos. The snow was so deep nobody could go out; they knew the Nazis would see their footprints, find them, and shoot them. They cooked potatoes in the bunker. They survived. Then the word came: *The Nazis are hunting for partisans. They are coming.*

The group abandoned the bunker. They lived in the woods and slept in barns among the horses and cows, slipping in after dark and leaving before dawn so they would not be seen. Some Jews, they'd heard, would pay the local farmers to let them stay for a night or two, to be hidden and fed and have a bed. But the farmers took only gold. Dora and Izzy didn't have any gold.

There was a Polish farmer who had known Dora's family before the war. The ragged band went to him and begged. *It's wintertime,* they said, *we have no place to go.* He agreed to keep them through the winter and dug a *ziemlanka,* a shallow shelter sunk in the ground, usually used to house animals. Most days, he would bring out food. And then one night he came empty-handed and called them out of their hiding hole.

"Let's go to my house," he said. "I have prepared a dinner for you, for all of you."

Freezing, emaciated, and exhausted, they were overjoyed to leave their hiding place and enter the warmth of a real home for the first time in months. The group sat around the table and began to eat together. The warmth, food, and feelings of closeness brought back memories of the large Friday dinners they enjoyed before the war, which had already vanished into some distant past. Although none of their families were wealthy, their lives had been rich with love, kindness, and communal joy. For a few minutes, they felt those lives of abundant connection that had been torn away from them resume.

Then it all happened so fast.

The door opened, men with guns started firing, and before Dora knew it, the room was plunged into darkness. More shots rang out as she scrambled around on the floor, desperate to get away. Later she would learn that the armed men were from an anti-Jewish Polish militia and that one of the Jews in their group had extinguished the light on purpose, grabbed his gun, and shot the militia's leader. In the rush of adrenaline, the men in Dora's group clambered out the windows while she and her sister crawled under a large oven where the chickens roosted in the winter. They held hands. They held their breath.

Later that night, in the darkness, they slipped out the window. When they finally found their men, they discovered that Izzy's head had been split open by a blow from a rifle butt. His face was covered in blood, but he was alive. The blood, Dora saw, had frozen on his skin.

They crept through the winter fields to another farm, another barn, another potential enemy they'd have to beg for shelter—another desperate risk, to trust someone again. But all they could do was pray that they would not be betrayed—and survive.

The story of Dora and her family's ambush may sound like a scene out of the latest Hollywood World War II blockbuster, but it's not. It's the story of ordinary people dealing with extraordinary circumstances that I knew intimately well.

Dora and Izzy were my grandparents. After the war, they immigrated penniless to Lithuania, where they married and where Dora gave birth to my mother. Then they moved to Israel and finally to Canarsie, in Brooklyn, where I grew up.

My grandmother and I shared a special bond. She lived a few dozen blocks from my childhood home and cared for me after school. I remember how she would sit on her porch and watch me ride my bicycle up and down the sidewalk, warning me not to stray too far. She was tiny—four feet nine on a good day. But she was a force of nature. She wore bright red lipstick every day and came after you with what I called "get away from me" kisses, landing them on both cheeks no matter how hard you tried to dodge them. She did not suffer fools; one of her favorite sayings, if someone crossed her, was *That bastard!* She had no formal education but was wise in her ability to navigate the world. I suppose, given her past, she had to be.

And there was always food at her house—more than you could ever finish. Perhaps because she had experienced scarcity, she wanted me to experience abundance. When I came through the door after school, she'd have elaborate meals laid out, which she'd press and press me to eat (no surprise I was a portly child). To this day, I haven't eaten a Michelin-starred meal that could compare with the burst of sweetness from the caramelized onions hidden inside her matzo balls, which perfectly countered the saltiness of her homemade chicken broth, or her sweet and tangy pineapple-studded noodle pudding.

She was like a second mom. And she showed, every day, that she loved me. But she was no-nonsense—tough as nails. And she never ever talked about the war. Except for one day a year.

Every year, on a crisp Sunday in the fall, my mom would drag me from my soccer game, still dressed in my cleats and typically muddy uniform, to attend the Holocaust remembrance day gatherings my grandparents held with other survivors. That was where I first heard my grandmother speak of the time she spent living in the woods, going days without food, surviving the winter in a thin dress and coat. It's where I heard her talk about learning that her mother, grandmother, and younger sister had been massacred in a ditch off the town square, and the moment she realized that her father's rushed farewell from the home where they were hiding would be the last words she ever heard from him.

Sitting in the hot synagogue where the gatherings took place, on stiff wooden benches I was convinced were designed by Jewish law to produce discomfort, I'd imagine how I would feel if it were me: the terror of my unknown fate, the anxiety of being hunted, the boiling anger over being betrayed, the unimaginable grief over lost loved ones.

Not exactly how a child hopes to wrap up the weekend after playing soccer with his friends.

But the thing that truly astonished me were my grandmother's tears. Aside from that day, I never saw her cry or express sadness. But one day each year, all the emotion came spilling out. And as the story came out, so did the tears. My grandparents sobbed, even wailed. These were people I normally knew to be pillars of stability, which made this display of raw emotion even more jarring.

Hearing my grandparents' stories filled me with questions: How were they able to endure atrocities and survive to have a normal, happy life, when so many others hadn't? How did they cope with the trauma of what had happened to them? And why did they keep all these emotions bottled up, all year long, except on this *one* day?

One night over dinner when I was in middle school, I finally blurted out my pent-up curiosity.

"Bubby," I said, using the tender Yiddish name for grandmother. "Why don't you ever talk to me about what happened during the war?"

My grandmother met my question with a long silence. An unusual stillness descended on the kitchen while I stared at her over the table awaiting a response. Finally, she answered. "E-tan," she said, unable to pronounce my name properly with her heavy Eastern European accent. "Darling, don't ask 'why.' "

It was a variation on what she said anytime I pestered her with questions about the war. "Why is a crooked letter" was one of her favorite rejoinders.

My grandmother barely spoke English, and yet she'd somehow learned this peculiar bit of idiom and made it her personal mantra. I understood what she meant: Sometimes there are no answers, and searching for them only creates suffering. "Why" is just a source of pain. Leave it alone, finish your homework, ride your bike with your friends. Cherish the life you have.

But my questions only multiplied. As I grew up, I became an observer of emotion. I began to question not just my grandparents' experiences struggling with their emotions but other people's more common experiences grappling with the emotional obstacles they faced, including my own.

Why did my dad, normally so patient and empathic, become a raging lunatic on the roads of New York City (to the enormous fear and embarrassment of his son riding in the back seat)? Was his obsession with transcendental meditation a tool he happened on to provide an antidote to the uncontrollability of his temper while driving?

Why did my friend Amy keep focusing—over and over and over—on whether she'd get into her top college choice if it only made her more upset? And why did she think having the same conversation with me about her anxiety every time we talked would help?

Why was I unflappable on the soccer field but overcome with stomach pains before asking girls out on dates in high school, dependent on my dad and friends for pep talks to help me muster the nerve to dial their numbers?

In college, I watched charismatic close friends who were star students in high school resort to alcohol and drugs to manage feeling like impostors during their freshman year. I witnessed my atheist relative surrender to religion to cope with his overwhelming grief at the loss of his wife.

It seemed as if we were all just stumbling along, occasionally finding an accidental or Band-Aid solution to help us manage our emotional lives. Sometimes our improvised tools helped. Sometimes they made things worse. It seemed so haphazard, isolating, and inefficient.

What does a person do when faced with questions like these? My grandmother's advice all those years ago was simple: Stay away. *Don't* ask why.

Fast-forward thirty-five years. I'm now a professor at the University of Michigan, where I founded and direct the Emotion and Self-Control Laboratory, a lab that specializes in asking "why" questions about emotions.

I can only imagine my grandmother shaking her head with amused exasperation.

Until relatively recently, the scientific establishment took the same attitude toward emotion that my grandmother once had: Mainstream researchers have long considered feelings, moods, and other emotional processes a confounding black box that was impossible to measure, and generally not worthy of "serious" study. But during the years since I first became interested in understanding people's struggles with emotion, the field of psychology underwent tectonic shifts. The territory of emotion, once marginalized and under-studied, has evolved into a vibrant area of research that draws scientists from multiple disciplines to shed light on many of the same questions about emotions that I found myself asking as a kid. We now possess a vastly improved understanding of the science of emotions—a science that teaches us about what emotions are, why we have them, and, most important from my perspective, how we can manage them effectively.

Speaking to these vast, universal, and timeless issues is the purpose of this book. Because as long as humans have walked this planet, we've struggled—mightily—with our emotions. Early human writing samples attest to this: three-thousand-year-old clay tablets discovered in present-day Syria and Iraq describe the pain associated with emotional states such as anxiety, depression, and heartbreak. Many of the tools we've discovered to deal with our emotional struggles haven't exactly stood the test of time. In fact, some of them are downright chilling to contemplate.

In the mid 1860s, an American diplomat traveling in Peru came across a remarkable artifact. Ephraim George Squier—also an archaeologist—paid a visit to a socialite who collected ancient artifacts. She welcomed Squier into her home and allowed him to inspect her Incan treasures. After admiring the myriad stone figures, sculptures, and other items she had accumulated, he noticed a peculiar specimen: a skull fragment that had been unearthed from an Incan cemetery.

Ancient skulls are common archaeological finds, but this was no ordinary relic. A nearly symmetrical half-inch square with clean edges had been carved out of the frontal bone, the area of the skull that sits above the eyes and encases the prefrontal cortex—the part of the brain that enables us to plan, manage our lives, and reason logically. Plenty of ancient skulls had, of course, been unearthed with damage, but usually that breakage was irregularly shaped and most likely the consequence of a traumatic event or prolonged exposure to the elements. The four surgically precise incisions on this Incan skull told a different story.

Squier sent the skull across the Atlantic to be examined by the famous French surgeon Dr. Paul Broca, one of the world's foremost experts on ancient human skulls. After inspecting the cranium, Broca concluded that the square-shaped hole represented clear evidence of a surgical intervention that predated the European conquest of the region in the sixteenth century and was performed on the deceased person *while they were still alive.*

The hole carved into the Incan skull resulted from a procedure that is now widely considered one of the first surgical techniques developed in human history: *trepanation,* or the creation of holes in people's skulls. That our distant ancestors were capable of carefully trepanning people's skulls is remarkable. But even more incredible is one of the reasons *why* they are believed to have used this intervention: to help people manage their *emotions*.

Consider that for a moment: One of the earliest forms of surgery in the history of medicine was used to help people regulate their feelings.

It is impossible to know exactly which emotional maladies warranted receiving a hole in the head thousands of years ago. Historians suspect the technique was likely used to help people manage extreme cases of depression, mania, and other conditions characterized by emotion dysregulation. Regardless, what we can say for certain is that carving a hole in a person's head to provide them with emotional relief was *not* a great idea. But when you look at the history of how our species has dealt with emotions since those early days of head carving, you see that the struggle has always been real. And for as long as we've been grappling with our emotions, we've been trying to find tools to regulate them.

Leeches.

Exorcisms.

Witch burnings.

We've come up with remarkably creative (and cruel) measures to control our emotions. In the seventeenth century, pressing a burning iron rod to the skull was a recommended solution for managing heartbreak, while centuries later mineral water was pedaled as a stress tonic. And as shocking as the treatment may seem to us today, the spirit of trepanation lived on until only a few decades ago in the form of the lobotomy, in which a surgeon would flip open the eyelid, slip an ice-pick-shaped instrument through a person's eye socket, and poke around in their frontal cortex to sever key neural connections. In fact, António

Egas Moniz, a Portuguese neurologist, shared the Nobel Prize in 1949 for developing the procedure to treat extreme emotional states. The structure of DNA, the discovery of insulin, the development of MRI technology—Moniz's procedure reached the same level of distinction as these extraordinary discoveries. We human beings have considered emotions so perplexing—so destructive—that we have resorted to carving holes in our heads, ingesting heavy metals, and disabling parts of our brains just to get some relief.

And—like our ancient forebears—we remain in trouble in the emotion department.

College campuses are overrun with students who require extra support to help with their emotions. Britain and Japan have ministers of loneliness, while the U.S. surgeon general has made combating social isolation a national crusade. Corporations invest millions in programs to address burnout. Even "the Boss" Bruce Springsteen himself has talked about his struggles with depression. We install apps on our phones to keep our stress levels in check. We spend money we don't have in a wellness industry that promises to make us just a little bit happier. A 2020 report found that approximately one in eight adults in America took an antidepressant every day to manage their emotions. A medication that, although genuinely helpful for many people, is far from a panacea.

Interventions have undoubtedly improved since the days of leeches and lobotomies. Our methods have become much subtler—and less damaging. Advances in talk therapy, innovations in psychopharmacology, and the blending of ancient and modern contemplative traditions have expanded our access to relief from emotional distress. And yet despite all these efforts, the statistics on mental health and well-being are going in the wrong direction and keep getting worse. More than half a billion people worldwide suffer from some form of depression and anxiety, afflictions that cost the global economy a staggering one trillion dollars a year.

One *trillion* dollars.

Patchwork solutions reside in myriad places, from the bowels of the internet to the dusty shelves of the library. As a result, many of us resort to cobbling together coping tactics that range from kind of effective to actually harmful without really understanding how they all fit together to help us (or not). A little meditation here, a cold plunge there, some cognitive restructuring, maybe a cocktail or five to smooth things out.

Meanwhile, people who *are* good at managing their emotions are less lonely, maintain more fulfilling social relationships, and are more satisfied with their lives. They experience fewer financial hardships, commit less crimes, and perform better at school and work. And they're physically healthier as well: they move faster, look more youthful in photographs, biologically age less quickly, and live longer. Simply put, the ability to control your emotions isn't just about avoiding the dark side of life; it's about enriching the positive, generative, and rewarding dimensions of existence as well.

The question we now face is the same one that likely inspired our ancestors to drill holes in their heads when they were struggling: What do we do about all these feelings?

In 2021, I published my first book, *Chatter: The Voice in Our Head, Why It Matters, and How to Harness It*. Its central question: Why do our attempts to work through our negative feelings often backfire, leading us to feel worse, and what does science teach us about how to harness our capacity for self-reflection?

Immediately after publishing the book, I set off on an extended book tour that lasted about two years. After events, people would come up to me wanting to talk. They were grateful for the book and had feedback about how it helped them, which was wonderful to hear. But they had questions—lots of them—about the broader terrain of emotions and how to manage them:

Should I always be in the moment?

Can you really control emotions?

Why do I struggle to do the things I know I should do in the heat of the moment?

It was as though I had just taught them about heart disease, which was great, but they also needed help dealing with inflammation, diabetes, and cancer.

People talked about having parents and bosses who brushed their feelings aside. They struggled to define what an emotion was and asked me why they never learned how to manage them growing up. I am not exaggerating when I say that some people would come up to me with these questions in tears. And it was *everybody*: elite athletes, CEOs, parents of teenagers, special forces operators, you name it.

What became clear to me during those moments was how curious people were about their emotional lives and how motivated they were to manage them. Which is why I decided to write this book. *To provide you with a blueprint for understanding your emotions—what they are, why they matter, and how to harness them.*

Our emotions, both the positive and the negative ones, are tools we use to navigate the world. They influence who we fall in love with and who we hate. They motivate us to stay longer hours at work to realize our dreams and rein in our aspirations when they no longer serve us. They can fill our lives with health and vitality or sap our energy when we lose sight of what matters. And they are often the difference maker when it comes to sustaining close bonds with others or becoming mired in relationships fraught with conflict. And yet, for all their profound impact on our lives, few of us receive a science-based guide for how to turn the volume on our emotions up or down, or how to transition gracefully from one emotional state into another.

Over the past twenty years, we have witnessed a scientific renaissance in our understanding of these mechanics of emotions—the psychological and neural "nuts and bolts" that explain how emotions work—and an explosion of new methods to measure and test those mechanics. Neuroscience techniques allow us to visualize where and how quickly different brain networks become activated when people experience and control their emotions. Smartphones and wearables enable us to observe people's emotional responses unfold organically in real time as they navigate the world. Internet technologies permit us to run experiments on vast numbers of people scattered across the globe and gather unprecedented amounts of data.

These innovations, combined with our more traditional methods of experimentation, have transformed our way of thinking about emotions. We've learned, for example, that the key to emotional salvation doesn't always involve living in the present, and that far from being toxic, negative emotions can sometimes help us in surprising ways.

And perhaps most important, we've discovered that there are *no one-size-fits-all solutions* to our emotional problems. Would you ever take your vehicle to the mechanic and expect them to fix it with one tool? Of course not! Cars are complex, dynamic machines that require different tools to manage different problems. The same concept applies to us. Our emotional needs change from situation to situation, person to person, and certainly day to day—even moment to moment. We need a wide range of tools to meet those needs. The great news: You already have them.

It starts with understanding what I call the major emotional "shifters" that we all have *inside* us. We can harness our senses to automatically shift our emotions. We can deploy our attention strategically in ways that help us conquer our greatest fears and savor our most joyous experiences. And we can alter our perspective on difficult circumstances to help us manage our most painful emotional states. These shifters help us move from one emotional state to another, allowing us to dampen or amplify feelings.

Each of these "internal shifters" can also be activated by forces *outside* us—the spaces we inhabit, the people we interact with, and the families, organizations, and cultural institutions we belong to. Understanding how these different "external shifters" affect us puts us in the position to make smart choices about how we interact with them to boost the power of the tools we have inside ourselves.

How and when to use these shifters, and which ones fit the job—that's what the science behind this book will help you figure out. Think of it as an instruction manual for the operating system you already have but might not be using to its fullest potential.

In the following pages, I'll share discoveries from my lab and others that have made these insights possible. You'll also read about powerful, surprising stories of emotion regulation from a cast of characters ranging from a happiness guru to a nuclear-code-carrying Navy SEAL with a soft side, to ordinary people like you and me stumbling into solutions that work (or don't). These stories highlight the ways that different shifters we all possess can be wielded to help you cope with what I genuinely believe is one of the greatest challenges we face as a species: how to manage our emotional lives.

The unconscious mind has a curious way of expressing itself.

Roughly twenty years after I posed that first "why" question to my grandmother and received her evasive yet caring response, I published my first scientific paper, a major milestone in an academic's career. Its title?

"When Asking 'Why' Does Not Hurt."

Taking a close look at our emotions doesn't have to be painful. It can be illuminating. Not only for improving our understanding of our own emotional lives, but also for helping others do the same. Our first step? Addressing the most basic questions of all: What are emotions anyway—and why are they often so painfully difficult to manage?

Part One

Welcome to Your Emotional Life

Chapter One

Why We Feel

Matt Maasdam sat cross-legged, hunched in a concrete box the size of a small sewer drain. He'd been there for almost two straight days.

Matt and his cohort had made it to the last forty-eight hours of the U.S. Navy's Survival, Evasion, Resistance, and Escape (SERE) School, and no one had eaten for five days. At first, Matt had been plagued by fantasies of the food and drinks he desperately craved—carne asada burritos from his favorite hole-in-the-wall in Coronado, cranberry orange muffins, and Mountain Dew. But those cravings had subsided. His body was now in survival mode, breaking down its own cells. This left Matt ruminating about what losing fifteen pounds of muscle mass would cost him next week, in the next phase of his punishing training, when he would be swimming in Alaska's icy waters. As a "new guy" in the SEAL teams, Matt was still proving himself, and knew he had to still somehow perform at his peak. He imagined pulling on his dry suit and feeling it loose against his skin from the weight loss, staring out at the two-mile swim ahead of him. He was a strong swimmer, but even in a dry suit hypothermia creeps up fast in thirty-degree water if you don't have enough muscle mass to insulate you. Every ounce helps. Not to mention that when sharing waters with sharks, killer whales, and squids the size of small cars, you definitely don't want to be any slower than usual.

As he sat crunched in his cell like a human accordion, his frustration began escalating. He could feel his heart racing and his jaw clenching. At times there was an angry buzz of electricity along his skin that made him want to start kicking at his cage, screw the consequences. The "guards" weren't allowed to beat the prisoners too badly, but they could use open-handed slaps, and if you made noise or called attention to yourself in any way, they could drag you out of your cell and bounce you off the concrete. Matt and his cohort had heard that a previous group of SEALs had broken out of their cells and taken over the camp, making a joke of the whole thing. They'd been warned—severely—not to attempt it.

SERE School wears you down; it's designed to. Plenty of people don't make it through the training. They either drop out or want to quit and get talked back in. There are the freezing desert nights; starvation; the physical torture of the claustrophobic cells; waterboarding; periodic beatings; and the loudspeaker playing a constant soundtrack of babies crying, tank treads grinding, boots marching, and Russians chattering.

It was a simulation, sure, but the scenario it was intended to prepare them for was very real. As SEALs in the field, they would face an enormous amount of risk and uncertainty. It was quite possible that one day they'd find themselves in this very same situation, except that everything would be real—the cell, the guards, and the physical violence they faced. They were being pushed to their physical and psychological limits, because if they broke, maybe they didn't belong.

It takes an unusual kind of person to survive SERE School. But Matt was determined to make it. No breaks. No tapping out. And frankly, he'd gone into the "prison cell" portion of the exercise hoping to learn something and thinking it would be relatively easy after going through SEAL training. Chill in a cell for a few days? *No problem. I got this.*

Now, the thing about being alone in a cell for two days is that you get closely acquainted with your own inner experience. Matt was acutely aware of seemingly every feeling and thought that arose as he sat, muscles burning, doing his damnedest to practice staying still, calm, and focused. There were blips of fear: *What if I pass out during the cold test?* There was frustration: *Get me the hell out of this microscopic box!* And mixed in with all that were warm feelings of happiness and excitement when he thought about the person in the cell across the prison courtyard from his.

Laura.

He'd noticed Laura during their first week of SERE School, when they were learning tactical sign language in a dank Vietnam-era classroom. When he glanced over at her, she glanced back, then held the gaze longer than was strictly friendly.

During the survival portion of the training, they'd been dumped in the high desert with a squad and instructed to survive for five days while being hunted down by the SERE School trainers. Hacking through the manzanita scrub, sucking on yucca for hydration, and crunching on ants for nutrients, Matt found himself scanning the horizon for her group; he had to keep reeling his thoughts in, reminding himself that he was out there to evade capture, not court Laura. But no matter how hard he tried to focus, she filled his thoughts. He picked handfuls of wildflowers and left them where he thought she might find them. One night, scanning for her, hopeful as always, he finally caught sight of her up on a ridgeline, silhouetted against the sunset. She must have spotted him too, because she gave a little mock curtsy before she disappeared.

Back in his cell, when the physical and mental discomfort was really getting to him, he'd lean over and look across at her cell, hoping to see her looking back. She had a unique capacity to make him laugh—even using sign language, not the most nuanced or poetic of mediums. The first day in their cells, she'd gestured around at her own concrete box and signed, *Four-star accommodations?* And he'd laughed out loud, risking the wrath of the guards.

Let's pause here for a moment and take a snapshot of Matt. He's in the middle of a career-defining training. He's starving. He's worrying about the next phase, even though there's nothing more he can do to prepare. He's got a bad crush. He's trying to stay focused and balanced—not tip too far into fear and frustration and, in the other direction, not totally lose himself in puppy love. In a sense, the most logical thing for Matt to do to get through SERE School successfully would have been to simply turn it all off; if his emotions had a valve he could have just cranked shut, wouldn't everything have been so much easier?

Most of us will never find ourselves in a concrete cell, battling through extreme survival training, but we share the urge to look for an off switch for emotions that sweep in at the most inconvenient times. On the one hand, we have to assume that we evolved to have emotions because they're helpful. On the other hand, they so often seem to be doing the opposite: sabotaging our health, undermining our ability to perform, and causing problems in our relationships.

When our emotions take hold, it can *feel* as if there were some kind of puppet master lurking inside us, yanking the strings. The way fear can paralyze us when we need to act, speak, perform; how anger incinerates our rationality when we so badly need to think clearly; how sadness can surge through the body like a crushingly heavy wave and spill over our edges when we desperately want to *not* show our private upset to the world. In these moments, it's hard to see how emotions are helpful. From that point of view, there's a logic to wanting to "turn off" our emotions at times. And if there's anybody who knows where that off switch might be, it would be a Navy SEAL, right?

Matt Maasdam does, in fact, have an emotional superpower. But it's *not* access to an off switch. And it has nothing to do with his elite level of training, his physical fitness, or his background as a Navy SEAL. It's simply this: He knows that success, in this moment or any, is *not* about turning his emotions off; it's about understanding how to *use* emotions skillfully without letting them completely take over. And that's a critically important insight because experiencing emotions for humans is like breathing air: Our emotions are both unavoidable and crucial to our survival.

Your (Very) Emotional Life

Emotions are not an infrequent experience; we are in a near-constant state of emotionality. Consider the results of a 2015 study that examined the anatomy of more than eleven thousand people's emotional lives for a little more than a month. Participants in the study strikingly reported experiencing one or more emotions *more than 90 percent of the time.*

Sometimes our emotions are so subtle we don't even notice them; they feel more like background music in the waiting room at the dentist's office. Other times, they become so sharp and deep that we truly believe we will never feel differently. And our emotions are not just contained inside our bodies. They tentacle out from us (regardless of whether we intend it) as we project, lash out, and exude curiosity or hatred or love. Sometimes we even "catch" other people's feelings, a phenomenon called emotional contagion, which happens online and offline and is exactly what it sounds like—emotions spreading quickly, person to person, like a virus.

Emotions are also an important fulcrum of action, and for better or worse they set us on our path, moment to moment. Every day, we are making decisions based on the flickering or flaring of emotions, and over time they influence the trajectory of our lives. How we feel about the people we love, the people we fear, or the causes we champion provides the fuel for nothing less than the greatest feats of human achievement. In 1631, for instance, Shah Jahan of India commissioned the Taj Mahal as a shrine to his beloved, who died giving birth to their child; Qin Shi Huang built the Great Wall of China for fear of nomads; and Harriet Tubman undertook risky missions to lead people desperately seeking freedom to safety via the Underground Railroad. There is a well-circulated quotation by Carl W. Buehner that illustrates why, when it comes time for a loved one to give your eulogy, emotions will be your legacy too: "They may forget what you said—but they will never forget how you made them *feel* [emphasis added]."

We don't typically experience emotions in a linear or compartmentalized way. They are varied, overlapping, and sometimes maddeningly confusing. In that 2015 study examining the anatomy of people's emotional lives, participants reported experiencing a mixture of positive and negative emotions *at the same time* around 33 percent of the time. One common example that will surprise no one was feeling love and anxiety simultaneously.

And our emotions don't only blend; they also interact. Sadness fuels anger. Joy makes grief more tender. If you've ever laughed at a funeral or cried at a wedding, you've experienced this complexity firsthand. In his cell at SERE School, Matt was pissed off and excited, full of joy and in pain. He was fearful about the future, and hopeful about it too. He was desperate to get out of that cell but also grateful to be sitting across from Laura. He was pissed off and stressed, but he was also falling in love.

So, welcome to your emotional life. It's a big, beautiful mess.

"What Is an Emotion?"

That's the first question I ask my students on day one of the course I teach on managing emotions. Usually only a handful of brave souls raise their hands. While some get close to a good working definition, almost no one is confident in their answer.

On the one hand, this is shocking: Emotions are central to our existence, yet no one is sure what they are? On the other hand, my students are not alone. For all the profound impact emotions have on our lives, and for as long as we've been studying them, there is no shortage of scientific theories about how emotions operate. The last time I counted, there were well over half a dozen.

Some scientists say emotions come in discrete categories, six or fifteen or twenty-seven emotions such as love, anger, disgust, and sadness. Others contend that our emotions come in a nearly infinite variety of colors, textures, and blends. Consider, for example, *Schadenfreude,* the German word for feeling good about someone else's misfortune. Some scientists argue fervently that emotions are universal responses that are deeply ingrained in specialized circuits housed in the biology of our brains. Others adopt a more contextual view, suggesting that emotions are an emergent property of how the brain responds to different circumstances. They believe universal emotion circuits simply don't exist, that every emotion we experience is utterly unique.

The bottom line is that my students are wise to be humble about their knowledge of emotions; there are many details concerning how they work that are still under investigation. But there are plenty of things in the world that we don't have neat and tidy explanations for—intelligence, cancer, quantum physics, to name just a few—that we nonetheless have learned a great deal about. And this is also true of emotion.

So, let's start with what practically everyone agrees on: Emotions are responses to experiences you have in the world, or that you imagine happening, that are meaningful to you; they're instruments to help you respond to the situation. For instance, if you see or imagine someone you care about fall off their bike, you can expect an emotional response. Emotions are triggered by things that are important to us—in this example, keeping people you love safe. This understanding of emotion is fundamental and generally tracks with our commonsense understanding of our own lived emotional experiences.

However, emotions are a lot more complex than that.

"Emotion" is an umbrella term that describes a loosely coordinated response that includes what we feel, think, and experience in our bodies in response to events we judge to be meaningful. You can think of it like an immune response. Some actions happen behind the scenes (T cells, B lymphocytes, and antibodies) while you are aware of the other consequences of that response (fever, chills, and sweat). The same is true of an emotional response; it has both conscious and unconscious elements. Here is how it unfolds. Once you encounter a meaningful situation, a cascade of responses is unleashed within you:

- *physiological reactions* involving your nervous system and other bodily processes

- *cognitive appraisals,* which reflect how you make sense of what is happening

- *outward-facing motor behaviors* that communicate how you're feeling to others, such as your facial expressions and vocal tone

This emotional response is flexible, which means sometimes one of the above processes goes before another and sometimes they work in parallel. They are also *loosely coordinated,* which means they tend to complement one another to bring about a certain function, although there are instances in which they can become decoupled, like when you can't judge what someone else is feeling from the look on their face.

What does this emotional response look like in real life?

Let's say you trip climbing the stairs to accept an award and experience embarrassment (no comment on whether this is an autobiographical example). In this situation you'd almost instantaneously—and unconsciously—experience a combination of physiological reactions that lead your heart rate to elevate, and appraisals designed to determine whether something like this has happened to you before and how you fared. Consciously, you might *feel* your face get warm, or the audience might see you grimace when you finally rise to your feet. Each of these reactions is part of your emotional response. Which component happens first and which ones operate simultaneously are matters of debate, as different theories emphasize different components. But it's certain that the emotions we experience are influenced by multiple distinct processes happening inside us for a purpose: to help us manage the many different situations we find ourselves in.

When it comes to emotions, there are two common sources of confusion. First, despite the cultural trope depicting emotions as the antithesis of rational thought, cognition—what we colloquially refer to as thinking—is actually a key building block of emotion. How we *think* about our circumstances shapes the emotions we experience; then those emotions reverberate back to influence how we think. For instance, if you walk into the SATs thinking you are bad at taking tests, your anxiety will be ramped up. Then you don't feel good about your performance on the test, and that becomes evidence for continuing to *think* that you're bad at test taking. In this way there's simply no pulling emotion and cognition apart. This bi-directionality of cognition and emotion allows us to modulate difficult emotions by changing the way we think. By thinking differently— *I get nervous sometimes, but I'm still a good test taker,* or *that jittery feeling is just excitement and anticipation, it means I'm ready*—you can work those pathways to your advantage.

A second source of confusion: the relationship between feelings and emotions. While thinking often gets pitted against emotions as if they were perpetually at war, *feelings* and emotions tend to be thought of as one and the same and used interchangeably. The truth is that feelings are simply the part of an emotional experience *that we are aware of.* And we are conscious of feelings in ways that we are not always conscious of in other aspects of our emotional experience (for example, an instinctive frown or shifting hormone levels). Feelings are like the "fever" of an emotional response, the conscious readout of what's going on behind the scenes.

Feelings are also a unique expression of our emotional experience, which is why no two people "feel" an emotion the same way. For some people anger feels like being under pressure, your insides ready to burst through your skin. For others, anger feels like hollowness—a dark hole in the middle of your chest. There are seemingly infinite ways that the different elements that make up our emotional reactions operate within us, which are influenced not only by the situations we encounter but also by our genes, environment, and personal history.

No matter how painful and overwhelming our emotions can sometimes be, it is essential to remember that we evolved our capacity to experience them for a reason: They help us navigate the world, which is why *all* emotions are functional, even the ones we don't like.

Bad Vibes: They're Not a Bug; They're a Feature

Good vibes only. Look on the bright side. Just change your perspective. Everything happens for a reason. It could be worse! Stay positive.

Part overcorrection to a mental health epidemic, part understandable impulse to distance ourselves from what feels bad, positivity is everywhere. But positivity taken to its extreme—and at the expense of hearing what our negative emotions have to say—can quickly have the opposite of its intended effect. Whether it's a solutions-focused work culture that makes employees hesitate to give constructive criticism or a well-meaning friend who urges you to forget about your failing grade and focus on your next vacation, the pursuit of positive emotions isn't always adaptive. Case in point: A 2013 study scrutinized the common practice of positive reframing. What the scientists leading the study discovered was that desperately seeking the silver lining could hurt *or* help, depending on the circumstances. When the problems bothering you aren't something you have a lot of control over (let's say you break a leg), it helps to reframe the situation positively. But when the sources of your stress *are* within your control (let's say your partner cheats on you again or you work in a toxic environment), looking for the silver lining is *harmful* and predicts greater levels of depression. When you can fix what's wrong—break up with the unfaithful partner or leave the demoralizing job—recasting the negative emotions into positive ones can prolong suffering.

While most of us have no problem reveling in emotions such as joy or excitement, we'd generally do almost anything to avoid negative emotions such as fear or shame. From that vantage point, negative emotions are the bad guy, so it's easy to believe avoiding them altogether is the key to happiness and success. But the fact is that our emotions—*all* of them—are a central adaptive feature of our lives.

Emotions aren't good or bad; they are just information.

There is an essential place in our lives for anger, sadness, guilt, grief, and a host of other "negative" emotions *when they're experienced in the right proportions.* The absolutist view that to live your best life, you need to rid yourself of negativity is a dangerous myth. Each of our emotions, however unpleasant in the moment, contains a powerful wisdom, shaped by evolution and experience.

You can think of negative emotions as remarkably sophisticated sets of software programs stored by evolution in the hardware of our physiology that help us achieve our goals.

Let's take a look at one of everyone's favorite bad-vibes scapegoats: anxiety.

In modern culture anxiety has become virtually synonymous with pathology. And while it's true that experiencing chronic anxiety is harmful, you wouldn't want to live life without this emotion. It is an elegantly adaptive solution for helping us deal with countless challenges. From the suspicious bear smell in the cave where our ancestors were about to bed down, to the email from our boss about impending layoffs, life *is* full of threats. Anxiety, like any emotion, can be maladaptive when it goes unchecked for too long (more about this later). But it has a fundamental adaptive function in that it helps us marshal a helpful response to either *approach* or *avoid* the threat we're dealing with so we can handle it.

Another culturally suspect feeling that few people welcome is sadness. It's what we experience in response to a loss we feel we can't replace—a missed opportunity for success at work, a friendship that can't be resuscitated, or the death of someone we hold dear. When sadness descends, we are bombarded with advice from friends, family, and coworkers about how to feel less sad—even if we're sad for a very good reason. As if any kind of momentary sorrow or wistfulness could lead to intractable depression. Yet experiencing and expressing sadness is useful. It slows us down physiologically in moments when we need to reflect, helping us take time to mourn a loss and shore up any remaining connections associated with that loss. And it also communicates to others that we need support. In fact, research shows that when people display facial expressions that convey sadness, others are more likely to help them than if they display an angry or neutral facial expression.

Beyond anxiety and sadness, there is a world of dark emotions that have a light side:

Envy can motivate us to work harder to obtain what we want.

Regret helps us avoid making the same mistakes twice.

Guilt guides us to recognize harm we caused and prompts us to make amends.

Anger can help us respond to a threat and correct an injustice.

Fear is a response to a specific danger that sharpens our awareness and compels us to act.

And lust can—well, let's not get into that, though it does help perpetuate the species.

In an experiment my colleague Aaron Weidman and I ran, we told participants to think about which emotions would be useful to feel in certain situations as they went about their day. For instance, feeling guilt for forgetting a friend's birthday might motivate making amends for the oversight, or feeling anxiety about making an error during a work presentation could contribute to more diligence in preparation. Likewise, feeling compassion when someone comes to you with a problem would help you support them more effectively.

Once our participants identified the helpful emotions—either negative or positive—we asked them to activate those emotions within themselves. This is easier to do than it might sound: Our logic was that all our participants had a lifetime of experience with emotions such as anxiety and excitement; we were simply asking them to recall those feelings and apply them to the current moment. What we discovered: People had no problem doing this with positive emotions, but they were highly resistant to *intentionally* feeling negative emotions. But when they allowed themselves to experience negative emotions in a useful context (for example, anger to correct an injustice, anxiety to deal with a looming threat), their outcomes were measurably better; they were more satisfied with how their circumstances resolved.

The point is, understanding that both good *and* bad vibes are part of a healthy emotional life gives us the capacity to accept and embrace our bad vibes with respect instead of trying to shove them away in panic. But there is, of course, a problem with emotions that we all know from a lifetime of experience: They are unwieldy.

Negative emotions are functional and adaptive, but also tricky to handle. The problem that toxic positivity overcorrects for is legitimate: In the same way that experiencing no sadness or anxiety would be detrimental to living a healthy life, experiencing too much of these emotions can have a profoundly negative effect. For far too many people this means mood disorders such as major depression and generalized anxiety disorder. But even for those who don't suffer from those conditions, when our emotions are not well calibrated, they can mislead us, undermine us, make us miserable, and cut us off from the very things we long for. When we become emotionally dysregulated for too long, the consequence isn't just a bad day.

Our emotional lives extend into our very cells. When we endure the physiological changes instigated by negative emotions for extended periods of time, they exert wear and tear on the body that contributes to a host of ailments—from an increased susceptibility to the flu and the common cold, to a heightened risk of serious illnesses such as cardiovascular disease and certain forms of cancer. And, it turns out, the negative implications of not being able to manage our emotions go far beyond these serious biological outcomes.

Insights from Down Under

In 1972, in the coastal city of Dunedin, New Zealand, a group of scientists began what should have been a small, short-lived study. Their mission: to explore the connection between birth circumstances and problems in child health and development. Little did they know they were embarking on a research odyssey that would last more than fifty years. Ultimately, their work provided a treasure trove of data that illuminated how our earliest years impact practically every consequential outcome that an intrepid set of scientists could imagine, from brain markers of cognitive decline to the quality of interpersonal relationships, to work success and beyond. The scope of the inquiry was both stunning and historic.

It began with the modest goal to check on childrens' development at age three, but evolved into a long-standing study tracking a cohort of 1,037 participants over time. Every few years the team carefully evaluated them on a wide array of physical, emotional, and psychological measures. The participants didn't just fill out the standard sets of surveys that are part and parcel of psychological research. They were put through a battery of tests that ranged from collecting sophisticated biometrics derived from their blood work to having 3-D scans of the interior of their mouths. They took cognitive tests, completed intelligence assessments, and participated in interviews in which they disclosed some of the most intimate details of their lives. To say the testing on these subjects was rigorous is putting it mildly. An article published in *Science,* one of the most prestigious scientific journals in the world, reported that "the intimacy of the data-gathering process make the group one of the most closely examined populations on Earth."

While the scope and length of the study alone make it stand out, what makes it truly unique was the way that the investigators meticulously measured the participants' ability to manage their emotions throughout childhood. Unlike most studies that rely on a single laboratory task or self-report survey to profile children's self-control, the project investigators ran the kids through multiple tests and asked their parents and teachers to rate them too. In so doing, they triangulated across several measures to get a precise and unprecedented window into how skilled these children were at regulating themselves. They performed these self-control assessments at multiple points over their childhood (at three, five, seven, nine, and eleven years old), which provided an excellent sense of how their abilities were changing as they grew. And then the researchers waited patiently, for years, to see what this childhood capacity predicted.

What they discovered was that participants' ability to manage their emotions predicted a lot about their lives.

Some findings were expected, like the positive link between early lack of emotion regulation and later substance abuse. But the vast range of connections went far beyond that. The kids who struggled the most with emotion regulation ended up becoming more likely to drop out of school and commit crimes. In contrast, kids who were adept at regulating their emotions advanced further in their careers, saved more money, planned more conscientiously for retirement, and were physically healthier. Remarkably, brain imaging scans and physiological assessments indicated that their brains and organs aged more slowly! Early childhood emotion regulation was so potent a factor in a person's development that it proved more influential than the socioeconomic circumstances of a child's family or even the child's intelligence levels in predicting several of these outcomes.

The Dunedin study findings suggest that our ability to regulate emotions exerts a powerful influence over the trajectory of our lives. However, there was another result hidden in the Dunedin study data that revealed an equally important truth: that no matter where we begin life in terms of our ability to manage our emotions, *we all possess the capacity to improve.*

Some of the children in the Dunedin study shifted how well they could manage their emotions from one assessment to the next. Some participants got better, and some got worse. And as kids' emotion regulation ability changed over time, so did how well they fared in life.

The results from this study highlight a critically important point: *Our ability to regulate our emotions isn't fixed. It is malleable.*

While there's no blood test that tells us how good we are

at regulating our emotions, there are signs we can watch for to alert us when things have gone awry. For instance, when it takes months to get a colleague's backhanded compliment out of our minds, or when we find ourselves exploding at our spouse over a dirty dish left in the sink. The problem is not that we feel hurt or anger; it's that we feel them too intensely and sometimes can't *stop* feeling those emotions long after we've gotten the message. Emotions that begin as functional can easily spiral into dysfunctional. These emotional misfires happen more often than we'd like and can usually be traced back to two key indicators; think of them as check-engine lights on your dashboard.

The first indicator: *intensity*.

Emotions guide our responses to situations by focusing our attention on the issue driving our reactions. But as anyone who has experienced an outsized emotional reaction knows, our emotions can overwhelm us, with our alarm bells ringing too loudly. When we explode with rage at our child's soccer coach for benching them or sink into a pit of despair when we can't fit into last year's jeans, we witness how our emotions can be out of proportion with our circumstances. Outbursts like these can damage our relationships, sully our reputations, and ultimately discredit our emotions themselves, causing us to mistrust or even fear them. Indeed, responding more intensely to day-to-day life experiences that trigger negative emotions predicts a higher risk of mental health disorders ten years in the future, as well as impoverished well-being over time.

The second indicator: *duration*.

Sometimes our negative responses don't subside, but instead linger. One of the key features used by mental health professionals to diagnose mental disorders is how long symptoms last, which makes duration a helpful marker for all of us. The tricky part is that our emotions are not light switches that we can just flip on and off, allowing us to focus, deal with the situation, and then move on. Instead, the duration of our emotions more closely resembles the amount of time it takes for a chemical compound to break down in the body, which varies depending on a host of factors.

One factor is the emotion itself. Sadness and hatred tend to be sticky, while shame passes away more quickly; a study published in 2015 found that sadness lasted 240 times longer than shame, which participants were quick to brush off.

The more meaningful an experience is, the longer the emotions associated with it last. Sadness, in general, tends to be about something that affects our worldview or identity, and this is one reason it's stickier than anger, which is usually about something specific. It's easy to forget your anger at the jerk who cut you off on the freeway because the emotion is centered on something fleeting that doesn't have much to do with you; it's just something that happened to you. Whereas sadness about the loss of a pet hangs around longer because your worldview has changed; the life you had with your beloved companion is markedly different from life without it. That said, if the anger is deep and intense—let's say about a betrayal—that too can change your worldview, causing it to stick with you much longer than anger about a mild transgression.

The amount of attention our emotions command also plays a role. If we find ourselves in the presence of the person driving our emotional response, it is likely to last longer because our attention is glued to it. Just ask anyone who has tried to be friends with someone they are still in love with!

How long our emotions last is also influenced by how we think about our experiences. Take just one facet of an emotional episode that we might be aware of: our physical response. Our bodies are often churning in ways that are apparent to us when we experience an emotion, and how we interpret those physical cues affects our experience. One experiment examined how people's interpretation of their physiological symptoms of stress affected their anxiety. It turned out that gently encouraging people to interpret their sweaty palms and racing pulse as signs that their body was rising to the occasion sped up how quickly they recovered from delivering a stressful speech compared with the people who did not receive any guidance on how to think about their symptoms.

Emotions rise in response to the circumstances in our lives in exquisite variations. Some flare up but sputter out quickly; others seem to linger for years like a song stuck in our head. No single factor determines how long our emotions last or how intense they will be. But here's the really good news: For all the ways we can't control the intensity and duration of our emotions, there are an equal number of ways that we can.

Mastering Your Emotional Stradivarius

Let's come back to Matt Maasdam, whom we left crouched in a tiny cell, in the middle of an endurance test designed to make him crack.

He didn't.

Matt made it through SERE training and he crushed the cold water training exercise in Alaska—the one he was concerned about. He excelled as a SEAL, parachuting into shark-infested waters, participating in countless missions, many of which you've likely read about in the news or seen portrayed in movies that I can't reveal here. He went to Harvard and earned a degree in leadership. And soon after, he was tapped to become a military aide to the president of the United States, carrying the nuclear football for President Obama.

There's a common notion that people like Matt—people who have the wherewithal to withstand SERE School, life as a Navy SEAL, and the pressure of carrying the world's nuclear future in their hands—have to operate in a Spock-like manner and repress their fear, desire, and anger. But when we unpack Matt's story, we can see that the opposite is true. Matt didn't excel because he was good at suppressing his emotions; he excelled because he understood them as signal, not noise.

Being able to alter the trajectory of your emotional response by speeding it up, slowing it down, or changing its intensity is called emotion regulation. And often people like Matt are excellent at it. But it's not because they don't feel things as deeply as the rest of us, or because they are better at shoving their negative emotions aside. On the contrary, Matt uses his emotions like an internal guidance system to help him navigate the world.

For Matt fear is a signal; it lets him know where to focus his attention. Anger tells him that there's a problem to be rectified. Likewise, when he experiences flickers of joy or curiosity, he gravitates toward them, knowing that he can reorient himself in their direction, often making a bad situation more tolerable.

In the last hours of SERE training, the prisoners were finally let out of their cells and given the task to rake rocks with their hands on the floor of the compound. Hamstrings on fire but free, finally, from his cell, Matt angled over toward Laura—eventually close enough that he could smell her hair, which he remembers smelling good despite the lack of showers. Soon, the American flag would be hoisted up, the national anthem blaring over the speakers to signal the end of the training. It was almost over.

"Hey," he whispered, finally able to get close enough to her to speak. "After this . . . do you want to go grab something to eat?"

And his future wife said yes.

Matt went into SERE training with an advantage: He knew that when used skillfully, our emotions—*all* of them—are critical guides through life's most consequential moments to some of our most inspired decisions. Even with extraordinary demands on his resources, Matt hadn't pushed away his emotions. But he also didn't allow them to spin out uncontrollably. He didn't ruminate about the conflicting feelings he was experiencing or the absurdity of courting a girl when he hadn't used toilet paper for more than a week. He was able to both feel the negative emotions *and* shift quickly back and forth between them and the positive ones generated by his interactions with Laura. If he hadn't been able to do that, he might still have made it through SERE training, but he likely wouldn't have walked away from that experience with the love of his life. And I wouldn't have the pleasure of standing in the parent pickup line at our local elementary school and seeing him wrap his son up in a bear hug when he gets out of class, Laura standing by his side.

———

Our emotions are our guides through life. They are the music and the magic, the indelible markers of our time on earth. The goal is not to run from negative emotions, or pursue only the feel-good ones, but to be able to *shift:* experience all of them, learn from all of them, and, when needed, move easily from one emotional state into another.

And that—like any skill worth having—is something that takes a bit of practice.

I look at our emotional life as a priceless instrument, a work of art capable of manifesting the near divine. Something a lot like a Stradivarius violin.

We are all born with an emotional apparatus—an instrument. But we've never been taught how to play it skillfully. How to position a bow, hold a long note, or recover when we flub a passage. Of course, like any instrument, we can play it poorly, making it screech so loudly that everyone around us covers their ears. Without learning the basics of how to play an instrument, how to interpret its sounds and manipulate it to our liking, we create noise.

Matt Maasdam is someone who knows how to play his emotional Stradivarius. Not only was he able to view his emotions as information instead of noise, but he also picked up tools along the way that helped him manage his emotions (we'll cover several of those later in the book). And yet, as exceptional as he is, Matt is not unique in his ability to manage his feelings. Through a combination of training, self-discovery, and genetic predisposition, he discovered how to wield a set of tools that we all have, but rarely use to their full potential. Most of us rely on intuition, cultural conditioning, and sheer luck when it comes to figuring out how to regulate our emotional lives. This is troubling, because if you aren't playing your emotional Stradivarius, you run the risk of it playing you.

And this is where we have a choice.

We have the capacity to play our instrument as beautifully as Matt. We all do.

The first step is realizing that you're capable of doing so in the first place. Unfortunately, as we'll see next, life throws countless experiences at us that seem to argue the very opposite.

Chapter Two

Can You *Really* Control Your Emotions?

When Luisa heard her daughter cough in the airplane seat beside her at thirty-five thousand feet in the air, she didn't think much of it. But by the second or third time, a cold recognition grabbed at her insides. She reached into her bag and found the granola bar Ella had just taken a bite of.

There it was, hidden under a plastic flap on the ingredients label: peanuts.

No, no, no, no, no.

Luisa looked over at Ella, who was watching *Daniel Tiger* on her tablet, and felt a kind of terror that she hadn't known existed. She knew there wasn't time to dwell on fear or self-judgment; she had to act. So, she did exactly what she had rehearsed in her mind a thousand times since learning of her daughter's allergy. She gave Ella a swig of Benadryl first and waited for other symptoms to emerge, hoping against hope that none would. No such luck. Within minutes Ella began writhing in pain and clutching her stomach. Then she began vomiting.

Luisa knew what she had to do.

Flooded with adrenaline, her hands shaking, she uncapped the EpiPen and peeled down Ella's jammies. Without hesitating, she jabbed her sweet girl in her tiny, chubby thigh and watched helplessly as Ella's eyes filled with pain and shock.

By then the commotion had attracted the attention of a flight attendant, so Luisa quietly gave her instructions to have an ambulance meet them at the gate. She looked at the flight path screen on the seat back and saw that they were only twenty minutes from touchdown. In most situations, this would be a relief, but she remembered reading that in bad cases anaphylaxis can be fatal in as little as twenty minutes. Feeling her heart begin to race, she tried to calm herself with the thought that if the EpiPen didn't work, there would be help waiting.

Thankfully, the epinephrine kicked in quickly; by the time they touched down on the tarmac, Ella's symptoms had vanished, and she was calmly reading a book. Ella had no understanding of the gravity of what had just happened, which Luisa was grateful for. But as a mother, she had just seen a nightmare come to life. She felt dizzy as they exited the plane and met the paramedics who cleared Ella to go home.

While those moments were the most terrifying of her life, for Luisa the aftermath was worse. In the moment, her own fear had taken a back seat to helping Ella; she'd been focused, driven, filled with purpose. But now she was overwhelmed by it, driven to distraction by incessant thoughts about what-if.

What if she hadn't packed the EpiPen?

What if it hadn't worked?

What if Ella had eaten the whole bar?

In the weeks that followed, Luisa was assailed by images of Ella dying. They came when she was washing the dishes and running out the door to go to the library. It seemed as if the smallest things triggered these thoughts—packing lunch, buying a new kind of cereal at the grocery store, or receiving a birthday party invitation. At first, the thoughts were about anaphylaxis. She would imagine Ella sitting next to a kid in preschool and snagging a bite of his delicious-looking peanut butter crackers while the teacher wasn't looking. Or she'd conjure an image of a well-meaning mom at a birthday party mistaking Ella's stomachache for overindulgence when it was really the onset of a fatal reaction.

Soon, thoughts of her daughter's death began to spiral well beyond peanut-related dangers. It was almost as if the fear were sensitizing her to all the other ways it's possible to lose a child. Uninvited thoughts of drunk drivers, teenagers texting behind the wheel, ungated pools, vicious dogs, and traumatic brain injuries unspooled like horror movies in her mind. And instead of sweeping the thoughts away, she let her imagination run wild, as if by rehearsing these potential catastrophes, she could prevent them. Knowing full well you couldn't cure an allergy, Luisa nonetheless began obsessively researching if there was anything she could do to better manage it. In the moment, the researching made her feel better. As if she were actually *doing* something to help. But the relief never lasted; the worry, guilt, and helplessness would surge up again, like a wave over her head. She couldn't seem to control her fears; they were just too big.

———

Luisa is far from alone in her feelings of helplessness, even if the thoughts that afflict each of us come in vastly different forms. Sometimes our emotions are rooted in a reasonable reaction to what happens to us. Other times it can feel as though our mind were hurling emotions at us from the window of a passing car like some sort of unhinged delivery driver.

While we can't change the fact that we have intense emotions, we *can* change our response to them through activating powerful neural systems that we all possess. When we access these systems, we have the ability to adjust the trajectory of our emotions—how long and how intensely we experience them. And that begins with understanding and leveraging a fundamental truth: There are things you can control and things you can't.

Emotions are both.

What We *Can't* Control

In the fall of 2000, a group of researchers asked 437 students who were about to transition from high school to college to fill out a questionnaire in which they were asked to rate their beliefs about these simple statements:

Everyone can learn to control their emotions.

If they want to, people can change the emotions that they have.

No matter how hard they try, people can't really change the emotions that they have.

The truth is, people have very little control over their emotions.

Almost 40 percent of the respondents believed that people *cannot control* their emotions.

As someone who has "Director of the Emotion and *Self-Control* Laboratory" in his email signature, this was a deflating finding when I first encountered it. Admittedly, there are understandable reasons that people such as Luisa and the students who participated in this study might think their emotions are running the show. After all, most aspects of our inner life *are* out of our control and even our awareness. We have hardwired reflexes for survival, like when we're startled by a loud noise or when we gag at a particularly noxious smell. We also develop learned responses that over time become almost as deeply ingrained as reflexes, like when I say "Love you, bye" to my children whenever they leave, no matter the circumstances. The same automaticity applies to much of our physiology. I've given thousands of successful presentations over the years, and yet, to this day, right before I go onstage, I find my stomach churning.

Our emotions have a similarly automatic quality. Let's say your favorite uncle cracks a good joke over Thanksgiving turkey. Naturally, you guffaw and spit out your mashed potatoes. The laugh is an expression of a spontaneous emotional response triggered by hearing something genuinely funny. The cascading internal processes that started with the punch line and ended with the laugh are invisible to you. You didn't hear the joke and think, *Oh, that was funny, I feel joy, the muscles in my face are beginning to contract into a smile, I should laugh now as an expression of that joy.* You just smiled and laughed. The entire emotional experience happened rapidly, and you had no conscious control over it. The same goes for reaching across the table to tenderly push a lock of hair out of your child's face. Or the way you walk faster when you hear footsteps behind you late at night in a parking garage. We can't help the joy we feel at a good joke, or the jolt of love we feel when we glance at our kids, or the fear that thrums through us in a vulnerable moment. These emotional experiences feel as though they just happen *to* us, because in a very real way they do.

We're also at the mercy of automatic thoughts that pop up outside our control. If you've ever experienced a dark or strange thought that seemingly came out of nowhere and caused you distress, you're familiar with this phenomenon; it's a common one. A 2014 study surveyed almost eight hundred people across six continents who were not suffering from any psychological disorders. They found that 94 *percent experienced at least one unwanted thought in the prior three months.* And these thoughts were only what they remembered and self-reported, which means there were likely many more. Scientists call these thoughts intrusive because they are spontaneous and often upset people. What do they look like? In one study a surprisingly large percentage of men and women reported periodically experiencing thoughts about driving a car off the road, having sex in public, insulting someone they don't know, or an intruder entering their home.

Just the other day I was at the gym carrying a dumbbell from one side of the room to the next when I suddenly imagined dropping it on the face of a middle-aged woman lying on a yoga mat off to the side. Did I want to hurt this woman? Of course not! It was just a thought I experienced, likely set off by a highly improbable fear that I would accidentally drop the weight.

Many loving parents picture themselves dropping their babies or watching helplessly as their toddlers run out into traffic. Luisa experienced thoughts like these about her daughter. Though they might have stemmed from those terrifying moments on the plane, they also went way beyond the reality of the incident. One idea why these thoughts bubble up into our awareness is that they are the brain's way of simulating worst-case scenarios to help prepare us for them.

That negative thoughts pop up out of nowhere, then, is not inherently problematic, and sometimes even helpful. However, it's what happens after these thoughts manifest that can really derail us. If the emotions they stoke—panic, fear, worry, shame—feed back into a loop of more intrusive thoughts, over time, that can contribute to impairments. Which is exactly what happened to Luisa. The intrusive images and thoughts related to her daughter's safety intensified the more she let them wash over her unimpeded. Over time, they became more frequent and distressing, eventually affecting her sleep, mood, and ability to be present with her family.

When you consider the prevalence of intrusive thoughts and feelings, and then combine those with the wild array of other automatic emotional experiences we have, it suddenly becomes easy to understand why 40 percent of those adolescents said they can't control their emotions. Because they're right. You can't. But—and this is a big but—they are right only if you focus on the first part of the emotional process: *the trigger.*

The trigger is the event that sets off our emotional reaction. A child in danger: *terror.* A traffic jam: *rage.* The scent of perfume that reminds you of someone long gone: *grief.* Triggers, of course, will be unique and personal for each of us—what sets me off will often be completely different from what presses your emotional buttons—and we can't control them any more than we can control any other aspect of the world around us. Triggers will happen, coupled immediately with our automatic emotional response.

We can't control the world around us. And we can't control the fact that emotions will arise. But that's only half of the emotional equation. A fire can spontaneously blaze to life, but once it does, we have opportunities to either extinguish it *or* fuel it. In other words, we can control its *trajectory*.

Emotions are not discrete flashes in the pan; they have unique lifespans. Anger at the rude checkout cashier might spark and die out quickly, while anger at a spouse because of infidelity might last for years. One day, the sadness about a parent's recent death might only lightly nibble around the edges of our minds, and the next day it might leave us flattened.

Emotions erupt within a set of particular conditions, and then, once those feelings manifest themselves in our awareness, we begin to change those conditions. For instance, when we keep thinking about a mistake at work, that focus (not the initial problem) is what causes feelings of anxiety to metastasize. The appearance of an emotion is merely the beginning: What we do or say or think affects the ongoing nature and timeline of the emotional reaction.

To understand how this unique machinery of control works, let's consider the *itch*.

What We *Can* Control

Once, on the heels of being aggressively assaulted by bedbugs during a stay in a tony New York City hotel, I made the mistake of trying to garden after returning home. Even though I had never so much as pulled a weed on our newly purchased property, I knew I needed a distraction. I was stressed out about the possibility of bringing the bedbugs back home to my family and at night was dreaming about insects crawling across my body, inflating in size as they feasted on my blood. Stepping outside, I plunged into the green unknown without so much as a glove and started ripping out what looked to me like invasive plants. It was exhilarating. A wholesome and decisive cure for worries that plague the mind. *Attaboy, Ethan!* I felt free. Until the next day.

I woke up to find both my hands and my sensitive male bits covered in what looked like itchy hives. Of course, I thought, those damn bedbugs. But a trip to the dermatologist eventually revealed the real cause: my rapturous pulling of weeds (read: poison ivy). Which led me to an even sorrier state than I was in before I entered the backyard. I was itchy. Very, very itchy.

Because poison ivy oils spread easily on the skin, it's not hard to find yourself covered in the stuff, which I was (one trip to the bathroom after a bout in the garden and I was doomed). But I also knew that scratching my itches was a very, very bad idea. Scabs, scars, potential infections. Despite my agony, I'm proud to say, I did not scratch. While many people might overlook this small feat, it's actually something to marvel at and celebrate. Not my individual fortitude, of course, but the larger evolutionary boon to emotional control that it represents.

Let me explain. I once heard Jonathan Cohen, a Princeton neuroscientist, talk about how many animals scratch as a matter of course; it's a sensory response that we're born with. But not all animals that itch can suppress their immediate desire to scratch to achieve a longer-term goal. This capacity, which is often referred to as cognitive control, allows us to modulate automatic responses, think abstractly, divert our attention, and alternate our thinking between different priorities.

Cognitive control is what allows us to plan for the future

rather than spend all our savings *right now* on whatever the well-targeted ads of the internet want us to buy. It's what allows us to compare two ideas side by side and weigh their pros and cons. It enables us to inhibit thoughts and feelings that we don't want to intrude on us, like when we want to text with friends but instead need to focus on work emails. Cognitive control is the reason we, and not our ape relatives, not only built the Pyramids, but imagined them in the first place.

The neural pathways that are responsible for cognitive control are found mostly in the prefrontal cortex, which is the part of the brain that sits directly behind your forehead and acts as a headquarters for complex thinking. Out of all the animals in the world ours is the most developed. While some nonhuman primates possess a rudimentary ability to exercise cognitive control, none possesses this capacity to the extent that we do. The fact that humans don't *have* to just react—that we have neural systems inside us that allow us to control how we engage with the thoughts and feelings we experience and the emotions they give rise to—is one of our species' greatest gifts.

So why did we evolve this remarkable capacity to control ourselves? There are two explanations that are popular, and they aren't mutually exclusive. According to the social intelligence hypothesis we evolved big, cognitively complex brains capable of control because of the advantage they confer for helping us navigate the challenges of social life. For instance, many of our cavewomen and -men forebears might have wanted to lash out in anger at someone important in the group, but those who survived were the ones who could manage their emotions, think ahead, and not get exiled. If you've ever tried coordinating a group dinner or, God forbid, a wedding, you know what it takes to juggle the emotions (yours and everyone else's) that erupt alongside personal politics, financial considerations, and logistical snafus.

The second explanation has to do with food. For our primitive ancestors, a steady food supply was not guaranteed. Survival demanded we remain flexible in environments that changed and didn't always present us with the nourishment we needed in the same place each time. Keeping track of what food was where and for how long, though seemingly a mundane activity, was and continues to be critical for our survival. While we may be a long way from the food shortages of our cave-dwelling days, I will admit that in our family, with our busy schedules, figuring out what to make for dinner, getting to the grocery store, and putting food on the table are certainly some of the most challenging tasks we face! The ecological intelligence hypothesis suggests that our need to manage this food-related uncertainty partly drove our development of cognitive control. We needed more sophisticated mental tools to coordinate those kinds of survival challenges. For instance, if you know food is not going to be plentiful at a certain time of year, you must be able to suppress the desire to eat everything all at once. You have to hold back, plan, and make strategic decisions that go against your desire to eat all the berries in cold storage.

Cognitive control helps us manage our emotions to solve problems, stay safe, and develop close bonds with others. And the story of how we evolved to do this illustrates something important: The ability to regulate our emotions is not a rare skill that only a few of us are lucky enough to have. It is a capacity that we all share, as a species. And it's what finally allowed Luisa to escape the emotional trap she was caught in.

One day, after a long period of struggling with intrusive worry and anxiety, Luisa stumbled upon a small but significant success: When a scary image of Ella getting hit by a car on her bike popped up, instead of doing what she usually did—letting the horror movie play out in her mind—she stopped the video. Or rather, Ella stopped the video when she came into the room howling over a scratched knee. Luisa got so caught up in helping her daughter that it wasn't until an hour later that she realized she wasn't replaying the intrusive image, she wasn't researching on the internet, and she wasn't anxious. Helping Ella had put a stop to the images that day. The cycle had been disrupted.

For the first time, she thought it might be possible to change the pattern. So, the next time a distressing automatic thought popped up, she stopped the video again—this time by *intentionally* distracting herself. And doing so didn't take much. It was just small stuff, like picking up a book or a magazine to redirect her attention. She also developed a technique where, if an intrusive thought popped up, she'd quickly reimagine it positively. So, for example, if she had a sudden image of Ella eating a peanut at a party and going into anaphylaxis, she'd flip it, and visualize her at that same party, but eating a hot dog and then leaping joyfully into the pool.

And once again, to her amazement, it worked.

Disrupting a cascade of negative emotions by distracting is only one way you can use your cognitive control system. Despite its name, cognitive control is not always activated to dampen emotion; we can just as easily use our cognitive control system to turn the volume *up* on our emotions as well. One classic study performed in 2004 put subjects in an fMRI machine, showed them gruesome pictures, and asked them to try to make themselves feel either better (that is, imagine the depicted situation improving) or worse (that is, imagine the depicted situation becoming more severe) at different points in the experiment. The participants were able to do this easily. But in both cases, the brain scans showed that elements of participants' cognitive control systems were activated, illustrating how we can flexibly use cognitive control to not only diminish our emotions but also amplify them. For instance, before a big presentation we might stand in front of the mirror and give ourselves a silent pep talk to turn up the volume on our confidence. This is cognitive control at its finest.

From the moment that a negative thought or unpleasant physical sensation enters your awareness, one decision point after another will present itself to you. This is the "trajectory" I was referring to after the initial trigger: opportunity stacked on opportunity to choose whether to elaborate, compartmentalize, or double down on that emotion. Every decision or lack of decision has implications for how an emotion will escalate (or de-escalate) to influence us, and then how that escalation (or de-escalation) will in turn act on our emotions further.

The ripple effect that our ability to control our emotions has on our well-being, health, and happiness is profound. And that is what haunts me about the study I mentioned earlier showing that 40 percent of adolescents don't think they can handle their emotions. If they don't believe they can manage them, will they even try?

You Gotta Believe

A geologist, a telephone repairman, a Peace Corps volunteer, and a golfer all walk into a lab in the late 1960s. What do they have in common?

Snakes.

More specifically, a debilitating, life-altering fear of snakes, or ophidiophobia.

This was the pre-internet era, so the renowned psychologist Albert Bandura recruited participants through an advertisement in the newspaper for his research. A smattering of humanity responded, describing the strikingly similar ways that ophidiophobia had derailed their lives. They'd given up outdoor activities such as hiking and camping and crossed gardening off their hobby lists for good. One participant shot himself by accident while trying to kill a snake. Many had recurrent nightmares about snakes, in some cases over the course of decades. In short, their lives were circumscribed by fear.

Hope for a cure was why the ophidiophobics answered Bandura's ad. Imagine their distress, then, when they arrived, met Bandura, and were promptly told that they would be entering a room with a snake in a cage. Recounting one interaction years later, Bandura said, "The first reaction is, 'This guy's off of his medication. I'm not going close to that room.' I said, 'Of course, because if you could, you wouldn't be here. I wouldn't ask you to do anything that you couldn't do with a little effort.'"

Once the participants' initial panic subsided, Bandura's research team asked them to perform a series of tasks that gradually brought them closer and closer to the creature they most feared. Researchers modeled the behavior and provided empathy and support along the way, never pushing the volunteers too far too fast. For instance, at first, they merely asked participants to look at the snake through a one-way mirror. Then they would ask subjects if they could move just three inches closer to the door, then six. The research team inquired about their fears—what specifically did they think was going to happen? In one case, a participant was terrified the snake would wrap around their neck. In response, a researcher showed the participant how he could put the snake around his neck without being harmed.

Through a series of supportive efforts, eventually the participant would be in the room with the snake. By the end of the four hours, they were holding the snake and marveling at its beauty. Years of crippling phobias evaporated in hours. As Bandura recalled, "The treatment not only enduringly eradicated phobic behavior, it also eliminated anxiety arousal, biochemical stress reactions, aversive ruminative thinking, and recurrent nightmares." One woman who had terrifying nightmares for years reported to Bandura a post-treatment dream in which a boa constrictor was kind enough to help her with the dishes (how considerate! I need one of those).

What was truly remarkable about the participants in Bandura's research was the way their newfound self-confidence rippled out, affecting *other* parts of their lives as well. Patients reported back to Bandura during follow-ups that when they ran up against a challenging situation, they felt better able to approach the problem with the confidence that they could solve it.

That was Bandura's lightbulb moment.

How did the snake intervention improve participants' ability to meet life's other challenges? Bandura was successful in curing the ophidiophobes in his experiment not only because he showed them their individual beliefs about snakes were wrong but because he also helped them discover the *mastery they had over their own emotions*—an ability long forgotten or perhaps never even known to them until now.

Bandura's historic research helped him uncover the power of a concept he would call "self-efficacy," the idea that if you believe you're capable of reaching a goal, that very belief plays a pivotal role in helping you reach it. It's not that the belief is some sort of magical elixir; it's that the belief itself puts you in a better position to make your goal a reality. A weak sense of self-efficacy—that is, a lack of confidence in one's own ability to create change in a particular area of one's life—leaves a person unable to help themselves, because they don't believe they can in the first place. In the case of the ophidiophobes, they spent years identifying with their fear, believing that it was more powerful than they were—that they could not change it. The result was that, without help, they didn't.

Take Luisa. At first, she didn't believe she could control the emotions that were triggered by each intrusive, peanut-laced thought. So, when triggered, she simply let the videotape in her mind keep playing, which exacerbated her negative thoughts and feelings. The trigger was determining the trajectory of her emotional response, and she had no idea she had the power to intervene.

Decades of research have revealed the power of self-efficacy to help or hinder us. The perception of our own self-efficacy is a "master belief," one that impacts what we are capable of handling in critical areas of life—like regulating our emotions. But these areas of impact go well beyond emotions. Roughly twenty years after Bandura's discovery, a group of researchers analyzed more than 114 studies involving close to twenty-two thousand people to evaluate the overall impact of self-efficacy on performance. The results? Self-efficacy was linked with a *28 percent increase* in performance, a significantly larger increase than many other established interventions that are commonly used, including feedback and behavior modification interventions. And it doesn't stop there. This belief in our own capacities shapes everything from whether we're able to moderate our alcohol intake, to how successfully we navigate relationship conflicts, to how quickly we recover from illness and injuries.

It's clear that this master belief is foundational, and it's important to get it right. So, short of hopping in a time machine and enrolling in Bandura's study, or hiring an elite coach, how do we improve our self-efficacy? How do we come to believe we can actually regulate our emotions?

Shaping the Arc of Our Emotions

We don't have to become snake charmers to improve our own sense of self-efficacy (though, clearly, it helps some people). Our beliefs about ourselves and how much we can control our emotions—no matter how ingrained they might feel right now—are malleable. Simply learning about what you *can't* control (automatically triggered emotions) and what you *can* control (the trajectory of those emotions once triggered) is critical for building your sense of self-efficacy.

We are all born with the capacity to manage our emotions, but most of us were never taught the specific strategies that allow us to step into the trajectory of our own emotional responses and change them. The next part of this book is devoted to introducing you to these tools and showing you where to find them, how they work, and how you can incorporate them into your life so you can use them skillfully to help yourself and those around you do the same.

To start, we'll talk about what is arguably one of the most primitive—and effortless—sets of emotion regulation tools we have. It's one that you use every day, but most often unconsciously, to turn down the dial on your anger or pump up the volume on your joy. If you've ever inhaled the scent of baking cookies and been transported to the warmth of your grandmother's kitchen, or listened to a song that hits you in the chest and reminds you of past heartbreak, you'll know what we're talking about next.

The senses.

Part Two

Shifting from the Inside Out

Chapter Three

What a 1980s Power Ballad Taught Me About Emotion: Sensory Shifters

It wasn't a particularly memorable fall Saturday morning, except for the fact that my younger daughter, Dani, was out of sorts. My family and I were about to pile into the car for her soccer game, normally one of the highlights of the week, but Dani was just not feeling it. All morning I had tried to figure out what was bothering her, partly because I wanted to help and partly because, frankly, she was bringing me down. Emotions are contagious, after all (more on that in a later chapter). My usual motivational tricks to shift her into a different headspace were failing miserably. Typically, Dani can't wait to get on the field. But when we finally made it into the car and I glanced in the rearview mirror, I saw her head flopped back on the headrest, her expression downcast.

Serendipitously, just a few minutes later one of my favorite Journey songs came on the radio. It wasn't long before I was pounding the steering wheel and belting out the chorus, admittedly with excessive enthusiasm: *Don't stop believing . . . hold on to that feeeeeeeling!* (No judgment, please.) In part, I was trying to shake the grump out of the back seat, but mostly I just really love that song. After a minute or so I looked back to see Dani bopping her head, smiling, and singing along. By the time we pulled up to the soccer field, she was ready. Opening her door, she tore out of the car, barely remembering to grab her cleats. A cheesy power ballad from the 1980s was able to do what no parental pep talk from me could—shift my daughter into an altogether different emotional gear.

As my wife and I hauled lawn chairs toward the soccer field and I watched Dani happily skip over to her teammates to join their pregame huddle, arms wrapped around each other as they psyched themselves up for the game, I thought about what a difference those four minutes and eleven seconds of Journey might have made for her that day. It was such a small thing: a song on the car radio. But it had completely reshaped her emotional state, perhaps the type of game she was about to have, and even, in the future, her memories of this day.

A memory of my own floated up: I'd just arrived at college at the University of Pennsylvania and felt painfully out of place on a campus swarming with preppy lacrosse players and BMW-driving international students. I was a middle-class Jewish kid from Brooklyn who talked like Robert De Niro and had an overfondness for sleeveless T-shirts. To make social matters worse, I'd been assigned to a remote dorm on the fringes of campus. I walked around during those first few weeks of the semester with a lonely, hollow feeling, wondering if I'd made a mistake and if I'd ever really belong.

One night, roaming campus with a friend I'd managed to

make—Anil, a fellow misfit, a first-generation Indian American from Tennessee with the thickest southern accent I'd ever heard—I came across a flyer advertising an a cappella concert. *A cappella?* At seventeen and coming from Brooklyn, I had no idea what that was. But we followed the signs across campus and into one of Penn's Gothic, castle-like halls, where a make-shift stage had been set up in a common room. Students were milling about and getting settled. It felt more like a dorm meeting than a concert of any kind. But when the music started, it was like nothing I'd ever heard.

The voices soared in the big room, with backup singers sim-ulating instruments using only their vocal cords, finger snaps, and claps. About halfway through the concert, an unassuming young student moved forward to the center of the stage and started singing Joan Osborne's "One of Us," and the energy in the room changed completely. The soulfulness radiating from his voice bathed us in warmth and excitement, seemingly bind-ing everyone in the room together with a palpable sense of con-nection.

We were transfixed.

Listening to him sing, I was no longer lost, homesick, fright-ened, or doubtful. I was exactly where I was supposed to be, following exactly the path I was meant to follow. Anil and I looked at each other.

Who the hell is this guy?

After that night, we went to their concerts every chance we had. Each time, we waited for the moment when that same singer, a guy by the name of John Stephens, would move to the front of the stage and transport us with his voice. After college, I learned he got a job at the Boston Consulting Group. Just an-other sharp Penn student with an extraordinary voice, I thought, making his way into the ranks of corporate life. You could imagine my surprise when I turned on the TV a couple of years later and heard the same voice that had moved me that night. There was John Stephens performing for a live audience, but now reborn with a new name: John Legend.

———

That was a quarter century ago. And yet I still think about that night, because a fleeting, momentary sensory experience rerouted my emotional trajectory. And maybe when she's my age, with kids of her own, Dani will still remember that car ride with Journey: the sun through the windows, the wind whipping her hair, the joy we both felt belting that song out together. These were, perhaps, small moments, but small moments become, in aggregate, the story of our lives.

I've been teaching about emotions for more than twenty years now. I know that calming sounds are effective at temporarily lowering blood pressure, that hearing birds twitter and chirp is linked with reductions in anxiety and paranoia, and that rats who have their sense of smell surgically removed show classic signs of depression. I've leveraged sensory experiences in my research time and again to alter the way people feel (which we need to do before we can learn how to help people manage their emotions). And yet, despite this knowledge and experience, if you had asked me as recently as a few years ago whether I purposely activate my senses to manage my emotions when I hit a slump, the answer would have been a resounding no.

After Dani's soccer game, I called a couple of colleagues and asked them if they had ever used music as a pick-me-up. The answer was an unequivocal yes across the board. Then I asked them if they knew of any leading emotion regulation frameworks outside of the domain of clinical interventions that used sensation as a tool to manage emotions. By emotion regulation framework, I'm referring to the models for how to regulate emotions that exist in mainstream science that we've built and tested. They include an array of tools—many of which are quite effective and which we will cover in this book—like situation change, attention deployment, cognitive shifts, and suppression. But the answer from my colleagues was a resounding no: None of them included sensation. So, what is a researcher to do when they encounter such a paradox?

Start digging, of course.

A Missing Link

Humans have been using sensory experiences such as music to influence emotions for millennia. Some evolutionary neuroscientists even believe that before we had command of speech as we know it today, we sang to ourselves and each other. It's one of the quickest ways to pluck an emotional thread: we sing to soothe our babies, turn up the volume on the car radio to feel more joy, and grieve through music at funerals. The effect music has on our emotions feels like magic; the right song in the right moment is like the sun coaxing a flower into bloom. The feelings that were hiding, the feelings that need to surface, the feelings we didn't even know we could express, unfurl.

Our ability to see, taste, touch, hear, and smell act as emotional levers. Much as we have been singing for millennia, we have also been engaging these other senses to manage our emotions. Take essential oils and aromatherapy. While this may seem like a new trend, the ancient Egyptians got into the scent game long before us (around 4500 BCE to be exact). One famous incense called *kyphi* consisted of more than ten ingredients including plant resin and raisins and was used to help people struggling with anxiety sleep. Shifting to the sense of touch, Egyptian healers were also apt to apply pressure to people's hands and feet as a medicinal intervention. In ancient India, meanwhile, Ayurvedic healers were mapping out six tastes that they believed could be used to promote inner harmony by balancing flavors such as sweet, salty, and sour. And when it comes to stimulating emotions via our eyeballs, we have been creating art to express our feelings at least since our prehistoric forebears in Indonesia were painting pigs on cave walls forty-five thousand years ago.

Our interest in changing our internal states through our senses hasn't waned since the days of the Pyramids, but it has certainly become more commercial. Businesses go to great lengths to leverage this connection—often outside our conscious awareness. Case in point: When my kids were younger, the first thing they'd say when traipsing through a hotel lobby is, "Why does it smell so good in here?" They were picking up on the fact that hotels often hire perfumers to create bespoke scents for their properties and then pipe those scents through the HVAC system hoping to induce brand loyalty in their unsuspecting customers. Coffee shops and restaurants commission murals and install local art on their walls with the aim of creating a warm and engaging experience for their customers. And legendary composers like Hans Zimmer play on our heartstrings through the big screen with musical scores that amplify drama and infuse it with subtle layers of emotional meaning. The pounding of drums signals an anxiety-provoking fight scene. The creaky synths and sinister guitars heighten our terror as we realize the ax murderer is already in the house.

Sensory experiences are such a foundational part of our everyday lives, and have been for so long, that it's easy to forget what an effect they have on us—I certainly did. And they don't have to be haphazard moments that randomly affect our emotions; they can be harnessed to *proactively* shift our emotional landscape. I didn't have to wait for Journey to come on that day. I could have seen my daughter's mood sliding precipitously downhill and queued up a playlist on my own. The irony is that while I was fully aware of how powerful our senses are for shifting our emotions, I still didn't use that information in the moments with my daughter before we got in the car, nor in countless moments before that in my life when I was the one who needed help regulating my emotions.

When my colleagues and I started combing through the research we quickly unearthed a 2011 study titled "Why Do We Listen to Music?" The researchers found that the primary reason 96 percent of 189 participants used music was to express their emotions and control their mood. Clearly, like me, these subjects were well aware of the link that exists between emotion and music. But if you ask people if they use music intentionally as a *tool* to regulate their emotions, you get a very different answer.

My colleague Micaela Rodriguez and I decided to run a series of studies at my lab at the University of Michigan, exploring the emotion regulation strategies of more than two thousand people. First we asked them to think back to a recent experience in which they felt angry, anxious, or sad. Then we asked them to describe the strategies they used to manage their feelings. All we wanted to know is which tools—if any—they used to turn the volume down on their negative emotions. Of all the diverse strategies the students named across the different studies, between just 10 and 30 percent of participants mentioned listening to music. A similar study by the University of Pennsylvania researcher Angela Duckworth and her colleagues found even starker results: Just 10 out of 577 high school students reported using music to deal with a situation that required self-control (the most common technique the kids said they used was cognitive reframing, which we'll turn to later).

It turns out there is plenty of research about how the senses affect our emotions, but very little on how to *strategically* use our senses to modulate our feelings. The more you look, the more you find a gap between what we know intuitively and what we put into practice. And this is where the opportunity lies.

Primitive Pathways

Like satellite dishes mounted to the exterior of cruise ships, our senses pick up critical information from the world that can be analyzed to help us navigate, steer clear of obstacles, and make crucial decisions. And they are powerful movers of emotion: The smell of a romantic partner's T-shirt can reduce cortisol levels and ease stress, as can petting a dog or hugging a teddy bear. We know that sugar activates dopaminergic pathways in your brain, and chocolate triggers pleasure. And in one seminal study, providing postoperative hospital patients with a view of nature hastened the pace at which they healed. These are small actions that can move our emotions with very little effort.

Here's how it works: The sensory cortex is one of the oldest parts of the brain. Located predominantly in the parietal lobe, the top back part of our heads, this hunk of brain matter is responsible for integrating information across our five different sensory modalities. The sensory cortex is exquisitely sensitive, which is important for discerning the difference between a blueberry and a toxic pokeweed berry or between a shout of joy and a shriek of terror.

At the simplest level, something in the world stimulates our eyes, ears, skin, tongue, or nose, and that information gets turned into electrical impulses that travel through our brain. Once these impulses hit the sensory cortex, the brain makes meaning out of the information it has received. Most of the time these sensory experiences are filed away quickly—the sound of wind through the trees overhead, the sight of the mail carrier walking down the street, the smell of the air freshener in your car. But sometimes, they are important enough to dwell on—like when we need to know whether to avoid the precarious dead tree limb overheard, or whether to approach the mail carrier with a complaint.

And that's where emotion comes in.

While countless sensations a day don't activate strong emotional responses, many do. There is an evolutionary reason for this: Emotions supercharge the meaning behind certain sensations we perceive to drive our behavior. In other words, when sensation activates an emotional response (fear of the tree limb hanging over your head), the sensory experience (the rattling of branches in the wind) is far more impactful. If the fear didn't arise, you'd be much less likely to avoid walking directly under that perilous branch.

So, let's say you hear the sound of the wind, followed quickly by the clunking of the tree branches. These signals are transmitted to the sensory cortex, and then in turn the sensory cortex fires off other neural signals to networks of brain regions that support emotional responses, which then help us figure out whether we should run or just gaze up in wonder. This sensory-emotion brain pathway has been critical to our survival for millennia; to dodge that lion or eat those mysterious-yet-necessary berries, your senses and your emotions must work together.

Another adaptive quality of the sensory-emotion pathway is its ability to help us remember the location of the lion's den and the bush that holds the nonpoisonous berries. After all, in order to learn from the sensory information that we gather, we have to remember it. This is the beauty of adaptation; instead of senses and emotions working together to help us survive just once, we developed a capacity that linked senses and emotions with memory, which helps us survive when the same situation arises in the future. Emotion is like the glue that holds the sensory information and the lesson learned together, so when that same sensory input floods our system next time, we know what to do. This is why the memories we form are stronger and reverberate in our awareness longer when they have an emotional tinge, often due to sensory experience. For instance, the smell of Marlboro cigarettes will forever conjure up memories of car rides with my dad, while a whiff of potato pancakes reminds me of holidays at my bubby's house no matter what time of year it is.

A well-known example of how the senses and emotional memories interact is an experience described in Marcel Proust's famous novel, *In Search of Lost Time*. In its opening pages the narrator is bowled over by an involuntary memory that springs up when he combines the crumbs of a madeleine cookie with a sip of tea. Dubbed the Proust effect, this phenomenon illustrates how deeply sensation-activated emotional memories can burrow into our brains. The book's narrator had not thought about mornings with his aunt Léonie for years, but when he bit into that buttery madeleine, the past experience of overwhelming joy came flooding back. The love and wonder of childhood, the bodily sensations of delight. The indescribable quality of innocence. The taste and smell of a cookie and the positive emotions of the past were forever fused together, accessible simply by re-creating the sensory experience.

What Proust tapped into in his novel is a sensory autobiographical memory pathway that is often the source of positive emotions. For instance, looking at old photographs, listening to Guns N' Roses, or eating my grandmother's matzo ball soup sends me directly down different wormholes of nostalgia. Unsurprisingly, research shows that these pleasant Proustian moments can reduce physiological markers of stress and increase positive emotion.

Naturally, the opposite can be true as well.

Unpleasant memories are also associated with sensations and can rocket you back into a distressing experience. For people who have been in car crashes, the smell of gas or burning plastic may trigger fear and stress. When smells or tastes trigger a negative reaction, this is known as the Garcia effect, named after a scientist who discovered that when mice were given a meal at the same time they were exposed to nausea-inducing radiation, in the future they would avoid that food. The remarkable part of this phenomenon is that this sensory-learned aversion can be developed after a *single* learning experience. Because cues to danger are so important, this one hairy situation gets filed away immediately to prepare you for a similar one in the future. This contrasts with most learning, which typically requires many experiences (think about how many times you have to listen to a song to remember the lyrics, or how long it took you to learn the movements of your favorite sport).

Sight can also send you careening into the past, which I

discovered early in my career while trying to construct an experiment to study the neural circuitry underlying people's experience of social pain. Our biggest head-scratcher was how to elicit intense social pain while also requiring people to lie flat in a cold dark tube also known as an MRI scanner. The solution we stumbled on was to ask participants to look at a picture of the person who recently dumped them and think about how they felt when they were dumped. As anyone who has ever been rejected knows, looking at old photos can bring all that pain rushing back, and it turned out to be quite an effective research strategy: We saw participants' social pain circuitry activate. But what was fascinating was that parts of the brain that code for physically painful sensations *on the body* were also activated by the social rejection memory. So not only did looking at the photo induce emotionally distressing experiences; it also had neural *physical* pain associations. The takeaway? When people say rejection "hurts," they may actually be referring to literal physical pain in their body. That's how strong this sensory pathway is.

When we understand the primitive and foundational nature of the relationship between emotion and sensation, it becomes easy to see why we take it for granted. It's hardwired within us—in the same way we don't consciously think about lifting our legs to run, feeling the breeze on our face, listening to the chirping of the birds returning in April, or smiling after that first slurp of a summer milkshake. We touch, taste, smell, look, listen. Then we experience fear, excitement, joy, or nostalgia. Coupling *sensation* with *emotion* just comes naturally to us. Which is one of the reasons it's so powerful. Another reason?

We're lazy.

The Law of Least Work

One well-known phenomenon that cuts across disciplines such as psychology, behavioral economics, and neuroscience is the law of least work—which means that all things being equal, organisms (including humans) tend to choose the path of least physical and mental effort. For example, if your goal is to reduce stress and you know that both listening to a song and writing in your journal for an hour will do the trick, which one do you think you'll choose? Good money is on the song because it is less physically effortful *and* doesn't require you to concentrate.

When it comes to mental effort, the law of least work operates because humans are "cognitive misers," which means we prefer actions that preserve our mental resources. As a result, we often unconsciously choose the path of least resistance rather than using our brainpower to seek out something new. Because of the volume of information that we take in on a minute-to-minute basis and the limited amount of resources we have to focus at any given moment, the more shortcuts the better. An example: Without even thinking about it, I take the same route to the grocery store every time I go. There are a handful of different ways I could get there. Yet, when I pull out of my driveway, I don't even think about it. It's not a conscious choice; my bias is to take the default route because I can drive it without having to focus. It has become a rutted groove in my brain that's nice and smooth. Taking a different path would require more mental friction. Effort, whether it is physical or cognitive, depletes our resources. So, if given the choice, our brain would like to just sit on the couch, eat chocolate, and binge-watch, thank you very much.

The law of least work helps us conserve critical resources,

sure, but it's a major obstacle when it comes to managing our emotions. Many of the strategies that we can use to suppress our reactions, to reframe a distressing situation, or to distract ourselves are effective but *effortful;* they take time and energy to successfully implement. This is one of the reasons people think self-control is so hard. Managing our emotions *is* often hard to do. But when we use the primitive pathways of sensation, we access a *relatively effortless* way to shift.

Professional swimmers like Michael Phelps are masters of conserving energy, both physical and mental. They will shave off all their body hair to gain a few milliseconds. Michael Phelps conserves cognitive energy by using a sensory experience to help him mentally prepare for competition. In the 2016 Olympic games in Rio, Phelps had a viral moment when he was caught on camera making a very intense facial expression as he was listening to the hip-hop artist Future's "Stick Talk." Phelps has often talked about the motivational impact of listening to music before a race, how it helps him relax, get in the zone, and perform at a higher level.

When you use your senses to shift emotions, you are piggybacking on ancient neural pathways that allow the act of shifting itself to be so easy it can even be unconscious. Sensation pushes emotion around well before the cognitive control system even engages. Person-to-person touch affects emotion within milliseconds, as does soothing self-touch, like when you rub your chin or temples. Our neurobiology is almost instantly changed by touch, bathing our brain in feel-good chemicals such as oxytocin and dopamine. Taste takes a little longer to process at around two hundred milliseconds, while fear-related triggers can jack into our brain network so quickly that we often aren't even aware of the sensory inputs. If a deer leaps into your headlights on a dark night, you've probably yanked the wheel before your mind has formed the word "deer"—and whether you've managed to avoid a wreck or have careened off the road is largely a matter of chance. In some brain imaging studies, emotional networks showed activation well *before* the subjects became aware they were even viewing scary images.

The speed of the sensation-emotion pathways is important because what sets sensory tools apart from many other emotion regulation tools is that they can change emotions *even when you're distracted or under stress* (which, of course, is when we often need to manage our emotions most). To state the obvious: For an emotion regulation tool to work, you have to actually use it. And that is where so many attempts to help people control their emotions fail, especially in moments of extreme negative emotion, which sap us of the very resources we need to use effortful tools. Talking to a trusted friend every day and journaling are wonderful strategies for shifting our emotions, and many people use them successfully. But plenty of people don't, because they take more effort and time than often feels feasible. Case in point: In 2020 my team administered a large study on coping with anxiety during the COVID pandemic in which we asked people to report on the tools they used each day to manage their anxiety. Journaling was one of the most effective strategies. It was also one of the most rarely used.

Using effortful emotion regulation tools is often effective, but they require time and concentration. Sensation, on the other hand, is like taking a neural bullet train instead. Studies show that if you drink something sweet, your body feels the emotional effects even if you're distracted. Touching your face during a memory task in which you have to remember the shape of something can be soothing. If your stomach is churning on your way to a first date, it's unlikely that you'll have the time or in-clination to pause and journal about your embarrassment. But if you can lower the volume on your anxiety by opening your phone and scrolling through some pictures of baby pandas? The pandas will win that competition.

Capitalizing on sensory experiences to shift your emotions is one of the easiest strategies for emotion regulation that you'll find in this book, and when it comes to shifting your emotions, ease is power. And yet, as with most good things in life, there is a shadow side to the ease with which our sensory apparatus can steer our emotions.

Sometimes, we unwittingly let our senses pull us in un-wanted directions. Research shows that when people are feeling bad, for example, they naturally lean into sensory experiences that perpetuate that negativity, a phenomenon called the emo-tional congruency effect. It reflects the reality that when many of us are feeling brokenhearted, we don't reflexively queue up Lizzo's "Good as Hell" to lift our spirits. Instead, we dive deep into Adele's backlist of moody, lovelorn ballads. This is not necessarily a bad thing. It's fine to want to roll around in your sadness at times. After all, as we've already discussed, when ex-perienced in the right proportions, negative emotions have their benefits. But being passively pushed around by the emotional congruency effect is *not* great if you'd really like to feel better.

Unharnessed sensory experiences can get us into trouble in other ways too: Who doesn't feel good with a mouthful of choco-late cake? Food is one of life's greatest pleasures. The same goes for sex. These are sensory experiences that we can use to punc-tuate our lives with more positive emotions. But like anything, overuse of these sensory experiences can become an emotional crutch that provides us with a short-term boost at the expense of long-term costs when we engage in chronic emotional eating or risky sexual behavior. What started out as a need for a little joy in a difficult moment can quickly turn into an emotional or physical problem. This underscores the need for us to *intentionally* harness our senses to turn the dial on our emotions up or down.

Sensation's Secret Side Doors

The last thing you'd think that Chayce Baldwin has time for is baking. He's a grad student in a demanding PhD program (mine), who works in a research lab (also mine), and he's the dad of two toddlers (adorable, but not my responsibility, thankfully—my kids sleep through the night now). This is a man who is often running on little sleep and a major time deficit. But when I asked him what his go-to strategy was when he was feeling stressed, his answer was immediate: "Bread."

Bread baking is not famous for being a fast process; even Chayce describes it as slow and meditative. But he says the sensory benefits of the experience hit immediately, the way the smell of coffee gives you a lift long before the caffeine enters your bloodstream. The minute his hands sink into the dough, his whirring thoughts and emotions start to settle.

"With sourdough, the live starter can differ from day to day," he said. "That means when I go into the kitchen to bake, I've gotta get my hands in there and pay close attention to how it feels; the stretch, the texture, the consistency, and all those things play into how I engage with it that day. The recipe changes every time, and you have to smell it, feel it, taste it. The senses are a huge part of it."

What Chayce is describing is a kind of sensory bundle—an immersive experience that hits multiple sensory channels at once. Within seconds of starting a new batch, he feels better—more grounded, more calm, more clear—which is ultimately what drives him into the kitchen, more so than the taste of warm, tangy bread that eventually comes out at the end, which is really just a bonus.

What stuck out to me about Chayce's answer to my question, though, was how aware and intentional he was. He knew baking bread helped him manage his emotions, he knew *why* it worked, and he was willing to drop everything and make the time, even on the most pressured of days. As I've watched his Instagram feed fill up with images of gorgeous crispy loaves, I've started to think of them as a "shifting diary," in a sense: every post marking a moment he decided to actively shape his emotional experience.

In retrospect, that moment all those years ago at the a cappella show was a wonderful experience *and* a lost opportunity. I left the concert feeling a distinct emotional shift. But I didn't take with me the new tool I had discovered. Instead, I just kept bumbling along, occasionally benefiting from sensory shifters for the next twenty years until I figured out how to use them more proactively. Now I see my car radio as an emotion regulation machine—a device that provides me with seemingly infinite options for changing the gradations of my mood. Instead of leaving this up to chance as I did as a lost freshman, we can use our senses strategically and skillfully to change our emotions.

The ability to shift our emotional life through sensation has been with us since we were embryos (touch is the first sense to be developed). We are all born with emotional side doors that can be opened to help us shift from one emotional state to the next. But what works for each of us is wildly variable: For you the local botanical garden might be the place you go to reshape your emotional state; for someone else it might be a monster truck rally and all the accompanying noise that serves to shift them out of the blues and into a more joyful state.

We talked about how all emotions—even the tough ones—have something to tell us and can be guides through life, and that's true. But we've also talked about how when they stick around too long, they can start to outlive their usefulness. That's certainly true of *loneliness*—something that's been so widespread even United States Surgeon General Vivek Murthy has referred to the "loneliness epidemic." Feelings of loneliness can certainly push us to connect with others, but when people have exhausted themselves trying, or are cut off from others for reasons beyond their control, that deep ache, that longing, can become physical, a suffering with no recourse. In a pilot study we ran asking people how they coped with feelings of loneliness when they were in that place of suffering, something we noticed was (a) how different everyone's answers were and (b) how several of them had gone *straight* to the senses. A few of the strategies they described:

- going into an ice cream shop to experience the nostalgic smell of waffle cones baking, evoking memories of summer days with friends (*scent*)

- putting his toes in the surf at the beach near his house and focusing on the rough sand and the pull of the waves, feeling that his problems were small in comparison to the vastness of the ocean (*touch*)

- to cope after the death of a parent, another reported looking up at the sky full of birds in flight: a sight that made her realize that she was a human being

experiencing all kinds of emotions, including
loneliness, which softened the sharpness of that
feeling (*sight*)

In a moment of deep loneliness, a colleague of mine described feeling so unsettled that he jumped in the car to drive
himself to his local beach with his surfboard: "On my drive
over, there is a steep road that goes downhill and suddenly reveals the sun setting above the ocean. . . . I put on some songs
from my favorite band, Sumbuck. I got to the beach. It was
golden hour. I got on my surfboard and caught a few waves. . . .
I didn't see anyone at the beach—I was completely alone. And
yet I felt connected. I didn't feel lonely at all." He'd thrown himself into an experience that hit all five senses: the sight of a
beautiful sunset, the smell and taste of the salt waves, the tactical sensation of flying over the water, the sound of waves—even
big, deep, sharp emotions can be washed out with that many
sensory channels open.

And as much as you can use this tool to push yourself quickly
toward feel-good emotions, you can also use it to explore the
more difficult ones—*at a time of your choosing.*

Anytime you've put on a sad song to cry along to when
you're alone, or a horror movie to have a little fun with fear in a
no-risk way, you've done this. Although research in this area remains ongoing, it's very possible that leaning into difficult or
complicated emotions in a safe space when we're ready and
open to them can help us work through loss and change, and
has the potential to help us feel less ambushed by those emotions when they pop up unexpectedly.

Knowing *which* senses, and *what* triggers for those senses

work for you, is the first step. And that's as easy as experimenting with the five sensory channels and figuring out which ones
give you the biggest, most immediate benefits without excessive
cost. So, before we go on, take a moment to consider the following questions:

1. Which one or two of the five senses pack the biggest
    punch?

2. And which comes with the lowest cost?

For me, the answer to question 1 would be taste and sound—
food and music are both powerful shifters for me—but question 2 helps me hone my go-to shifter, because for me leaning
too far into taste does come with a cost. The sensory delight of
dessert, I admit, has a potent effect on my emotional state: My
worries and cares melt away as soon as the chocolate peanut
butter cup hits my lips. But if I used this tool every time I felt
stressed, my health would suffer, undermining the whole point
of the tool. Taste (and chocolate peanut butter cups) are one
of the great pleasures of being alive, so I'd never deny myself
completely, but as my sensory shifter go-to music is the one. It's
immediate, it's effective, it's zero cost. I have a playlist on my
phone for different moods I want to shift into, and I use them
multiple times a week. And the key is that they're already there
on my phone, just a button tap away.

In the midst of big emotions, we're often blind to where
these doors are, or even that they exist. So I'd suggest identifying a few sensory shifters right now that you'd like to try using
intentionally and give them a test drive. Petting a dog, hugging

the warm body of someone you love (as long as it is mutually desired!), running your hand over a cedar branch to release the scent—these all have unique, individual effects on our nervous systems. So, do some experimenting and start collecting the ones that are your personal sensory prescription.

The real tool here is often just *noticing* these opportunities and focusing on the sensory experience of that task, instead of letting it slip by unnoticed. And look for those "sensory bundles," such as cooking or immersion in nature; experiences that braid more than one sensory channel together activate multiple sensory pathways in the brain, generating a multipronged experience. Which brings us to another shifter in our emotion regulation toolbox: *attention*.

Part of what we're doing when we tap into a sensory experience is diverting our attention *away* from feelings of stress, sadness, or worry. However, we don't need to rely on external sensations alone to divert our attention. Humans have developed a set of brain systems for steering our attention toward or away from different aspects of the world, and that includes our own emotions. To see how what we pay attention to can either help or hurt us when it comes to emotion regulation, we need to address the puzzle that's perplexed me since I was a child.

Chapter Four

The Myth of Universal Approach: Attention Shifters

In a video from the mid 1990s, an old woman's face fills the frame. It was filmed before the era of high-definition television, making the image resolution a little fuzzy to the modern eye. The piece of furniture over her right shoulder is hard to discern. It could be an ornate shelf, or maybe a lamp. She has short reddish-brown hair and warm brown eyes. Gold hoops with white baubles hang down from her ears to match her necklace and tasteful red suit jacket. Her deeply lined face fills with emotion as she describes the last moment that she saw her father, fifty-five years earlier.

The woman's neatly painted eyebrows knit together, and she leans forward as she repeats in a thick accent what he said to her: "My daughter, my child. Get dressed. They're killing already on the streets, they don't take like they used to take, you know, they're killing on the streets. Get dressed."

When she says the words "I never saw him again" to the interviewer, she doesn't break eye contact, but soon emotion overwhelms her. Her eyes drop to the floor, and she wipes her nose with her hand. Before she can go on with the story of how she escaped her father's fate, she dabs her eyes with a well-used tissue and takes a deep breath.

Her voice is still thick with grief when she continues the story: the screaming in the streets, the confusion about where to run to next, and finally the decision to hide in a giant box in the house she was sheltering in all day and through the night. It was a decision that would save her life and lead to innumerable other decisions, including the one fifty-five years later, to tell her story to a stranger so memories like hers wouldn't be lost to history.

In 1993, the Tower of Faces exhibit at the United States Holocaust Memorial Museum in Washington, D.C. had just gone up. I went with my family to see it; the story of my grandparents and their little lost town was featured. I still remember looking at the photos of her and my grandfather, and other young people from their town, most of whom had not survived. They looked out at us in black-and-white and sepia tones—typical young people captured on film before the war, before all the trauma that changed them. I was only thirteen, but I remember realizing the weight of it all: The people in those pictures were not so much older than me.

When I was researching this book, and pulled up the full, unabridged version of my grandmother's interview, I kept backing up the tape to those moments of profound emotion. She tells her story with a depth of sadness that wasn't frequently visible to me growing up ("Why" is a crooked letter, after all). Knowing her story as I do, I have struggled to understand how the kind of human suffering that she endured could be so effectively stoppered. In my experience of her, she was the archetypal Greatest Generation grandma—loving, strong-willed, and fanatical about cooking and taking care of her family. So where did all that emotion go? How did she manage to keep it locked up inside, and did she suffer for it?

The prevailing wisdom—from the therapeutic community to social media wellness influencers, and even to much of the mainstream research in the emotion regulation space—is that to deal with negative emotions, we must face them head-on. Until we deal with them, they'll stay inside us, growing more powerful as time goes on. *Avoidance,* we think, is unhealthy. Distracting yourself is, in the long term, a damaging practice. What the question really boils down to is this: When hard emotions surface, what should you do with your *attention*?

Attention is the mental spotlight that determines what information enters our awareness. Most of the time this spotlight is on autopilot. It whips around to whatever loud noise or scary intrusive thought pops up. But one of the qualities that distinguishes us from every other species on this planet is our capacity to consciously direct our attention wherever we choose. We can focus on a thought, feeling, or object for long periods of time when we want to, even when it's difficult. Likewise, if something unpleasant has our attention, we can swing the spotlight away and refuse to look at it if we don't want to. In one case, you are approaching the object with your spotlight; in the other, you're avoiding it. These are the two poles of attention that we toggle between constantly, and because we often do it automatically, we forget just how much it matters.

It's not always easy to point our attentional spotlight where we want, to keep it there without distraction or sometimes yank it away from something particularly salient. But we *can* direct and manipulate it, to varying degrees depending on the context, and what we give our attention to profoundly shapes our emotional experiences. We know this intuitively, even if we are not conscious of it. Let's say an intrusive worry pops into your mind about whether your upcoming flight is going to get canceled because of bad weather and derail your vacation. Focusing on that worry (refreshing the weather app every five minutes) or not (deciding to write that overdue work memo) has remarkably different emotional ramifications. You could go deeper into rumination, or you could brush off the intrusive thought, shift out of worry, and do something productive.

But in some situations, we don't have the luxury of immediately turning our spotlight away from what is distressing. Sometimes we just have to deal with what is in front of us—the car accident we just experienced, the fight we're having with our friend, or the alarming test results we just got back from the doctor. We can't look away in the moment because the concern requires immediate attention.

But what about the times when we can?

A significant part of our emotional lives isn't about what we're currently facing; it's about the aftermath of an event or anticipation of the future. The dramatic life events, yesterday's foibles, tomorrow's potential for rejection—these are the emotional experiences that frequently activate our hearts and minds. What do you do when you can't stop thinking about that mildly inappropriate joke you told last night in front of your boss's boss? What do you do when you want to fall to your knees in front of the refrigerator because a wave of grief over the death of a sibling just flooded your system? How do you keep moving through your day even when the emotional gremlins living in your head feel like they are tearing the place apart? Do you look away from your sadness about the end of a relationship and feel better, but at the risk of letting the despair rebound?

Unlike my grandmother, many people believe there's only one answer to these questions: *Confront your emotions.* It turns out both science and everyday life experience teach us that the answer is much more complex.

Walking into the Fire

One of the reasons I went to graduate school was to gain some clarity about the emotional forces that in so many ways define our time on earth. Imagine my excitement when a few weeks into classes I came across a classic paper titled "Emotional Processing of Fear" that gave me exactly what I was looking for: an elegant and simple framework for understanding what goes on in our minds when we have recurring negative experiences—and how to rid ourselves of them.

Edna Foa and Michael Kozak focused their paper on fear, but their ideas apply easily to other negative emotions that keep reverberating in awareness as well. A recurrent negative experience could run the gamut from the overwhelming terror you experience when you see a spider, to rumination about a perceived romantic failure. Foa and Kozak's framework for understanding these experiences is simple: Fear occurs when you activate "a mental representation" of something that you've interpreted as a threat. To put it simply, your brain has identified something that is scary and hooked it to an image, memory, or thought in your mind. When that threat-associated image, memory, or thought crops up, you feel fear. For instance, if you were in a car accident, you have memories of that experience, and every time those memories are activated (by a bad dream, an intrusive thought, a smell), you feel distress.

You can think about these mental representations as hard-to-watch videos of events that happened in the past, are happening in the present, or will happen in the future. When significant negative experiences happen to us, it often feels as if we can't stop thinking about them. We do this because our brain is trying to make meaning but has hit a snag and is trying to course correct. Maybe we did something that is out of sync with our sense of self ("If I'm a good person, how could I have hurt my friend like that?"). Maybe our sense of physical safety in our home is disrupted after a recent burglary ("I thought this environment wasn't a threat; now it is"). Maybe we feel socially rejected by people we thought cared about us ("What does it say about me that this person no longer wants to be my friend?"). We have systems in our minds dedicated to monitoring for these kinds of meaning-making snafus that shift into high gear when they are detected, pulling our attention toward resolving the issue. It's almost as if our mind keeps looping the tape and hitting the pause button, hoping we'll solve the puzzle so we can delete the tape.

What got me so excited about Foa and Kozak's framework was how elegantly—and simply—it illustrates the attentional crossroads we find ourselves at when we are struggling with recurrent negative experiences. Our brain is telling us to stop and fix this, and it won't let us rest until we do. So, we have to make a choice. Do we focus our attention on the nagging feeling that keeps popping up? Or do we move our attentional spotlight somewhere else and hope it goes away? Approach the problem (look at it, edit the tape) or avoid it (look away, distract yourself).

Foa and Kozak advocate for approaching negative experiences that you just can't shake. They point out that these representations will stick around until you intervene to rework them. Chronically avoiding them, according to their theory, allows them to linger and metastasize. So Foa and Kozak say to do the opposite: Face what you're scared of, and when you do, you'll see it's not as bad as you think. Approaching the feared object or experience repeatedly will update the mental representation, which will change the emotional experience and stop the looping tape. The idea is that the more you come in contact with the thing you are afraid of (for example, a spider), the more you'll learn that the outcome you are afraid of won't come to fruition (most spider bites won't send you to the hospital), and over time you learn that there's nothing to be afraid of. And as we saw with Bandura's snake experiment in chapter 2, you also become more self-efficacious, learning you can handle the difficult emotions related to the fear when they arise.

With all this in mind, it's easy to see why I was concerned about my grandmother. I knew that she had traumatic, harrowing experiences in her past. And I also knew that for the most part she didn't approach them. According to a preponderance of the research, approaching these negative experiences rather than avoiding them would have been helpful to her as a survivor of a major traumatic event. But why then would we have evolved the capacity to avoid in the first place? And why was my grandmother doing so well if avoidance is always so bad? Is it possible that sometimes avoidance can be helpful?

What Dennis Rodman and My Grandmother Had in Common

In June 1998, the Chicago Bulls were ahead of the Utah Jazz in the NBA Finals two games to one. If they won the series, it would be the team's second "three-peat" (three championships in a row) and a historic moment in basketball. Game four was in a few days, and the team led by Michael Jordan and Scottie Pippen had two days of practice and media events in the meantime. The Bulls were feeling confident, but there was a problem.

Dennis Rodman was missing.

The league's best rebounder and most famous bad boy had skipped practice and a mandatory press conference. Not to ice a sore ankle or take some family time, but to appear at a World Championship Wrestling match in Detroit sitting next to Hulk Hogan, smoking a cigar, and trash-talking.

If the world was shocked, the Bulls head coach, Phil Jackson, was not. Like everyone else, he knew all about Rodman's freewheeling life off the court: dating celebrities, like Madonna and Carmen Electra, throwing wild parties, dressing in a wedding gown and claiming to marry himself. He'd also been badgered by the press earlier that year when he'd skipped town mid-season for what was supposed to be a forty-eight-hour Las Vegas "vacation" that turned into a four-day escape. But after two successful championship seasons together, Jackson knew that Rodman's ability to maintain almost inhuman levels of hustle and tenacity on the court depended on his ability to get some distance from it all on occasion. As Rodman's teammate Steve Kerr put it, "Dennis was bizarre, but I think what made it work was Phil and Michael's understanding that to get the most out of him on the court, you had to give him some rope. And they gave him a lot of rope."

When I look at Rodman's antics, I see more than just a party boy shirking his responsibilities. I see someone strategically using distraction and avoidance to regulate their emotions. Rodman's determination to step away from the stress and anxiety of such a high-pressure position was an effective counterbalance to his focus and determination on game day. This was something that as a good coach Jackson understood. Reflecting on this years later, Rodman said, "I think Phil realized that I needed to always do me, just go do what I do. They're gonna get one hundred percent when I'm on the court."

Rodman's unusual approach to dealing with the intense pressure put on an NBA star worked for him. Not only did he continue to perform at a high level after his bender in Las Vegas, but in the fourth game of the Finals Rodman grabbed fourteen rebounds in just over twenty-nine minutes of play (for reference, fifteen rebounds on average *per game* made Rodman the leading league rebounder during the regular season). The Bulls won the fourth game and would go on to win their sixth NBA championship in eight years. Rodman had cemented his Hall of Fame career and remains one of the most famous defensive players of all time. And he did all this by doing what many people think is antithetical to effective regulation: building a healthy dose of avoidance into his demanding, high-profile career.

As for my grandmother, I'd puzzled over her apparent avoidance for years. I'd swing between worrying that she "bottled up" her emotions about that traumatic time in an unhealthy way and wondering if she was an aberration—a woman with a psychological superpower. But the septuagenarian that I watched describe her father's last words in the video was not a stoic. She was suffused with emotion, shot right back into the past, fully feeling the reverberations of her trauma. I wasn't wrong that my grandmother suppressed emotion in her day-to-day life. She certainly did. But what I didn't understand, at least not then, was that her superpower wasn't denial; it was her ability to *flexibly* deploy her attention to what she'd endured.

Watching that video of my grandmother at the Holocaust Museum led to a conversation with my mother in which I learned that my view of Bubby's emotional life was only partial. According to my mother, there was a lot about my grandmother I hadn't perceived as a child. I saw her as mostly avoidant. But according to my mother, my grandmother *would* talk about her experiences during the war to other people. My mother remembers them running into an old friend of my grandmother's in the grocery store one day. This friend, it turned out, had been in hiding with my grandmother in the Polish woods. On the car ride home from the store, Bubby talked to my mother at length about the experiences she'd been through with this friend during the war. It became clear that my grandmother hadn't completely shut down about that time, as it sometimes seemed from the outside; she just selectively engaged in those memories and emotions.

This broader understanding of my grandmother unlocked a mystery that I had unknowingly been trying to solve since my early days of gorging myself on potato latkes and noodle kugel in her kitchen. My grandmother was a person who had endured unimaginable trauma that had never been "dealt with" in any formal way (therapy was heavily stigmatized in the postwar culture my grandmother was raised in). And yet neither her harrowing experiences nor her lack of vocalization about the ordeal seemed to have a lasting detrimental impact on her life. For all the tumult of her twenties and thirties—escaping genocide, homelessness, forced migration, and poverty—my grandmother's life was bookended by happiness and fulfillment. She immigrated to the United States, nurtured her family, and became a contented person who just happened to have an extraordinarily difficult past.

My grandmother did not "face" her trauma the way many in society today consider healthy: no therapy, stages of grieving, or outward displays of being "in touch" with her feelings. She compartmentalized her grief and faced it sporadically. And by all accounts, that seemingly stoic, avoidant approach worked for her.

Meanwhile, the misperception that when it comes to emotions, avoidance is always toxic persists. I call it the *myth of universal approach*—the idea that we should always confront, express, and process our difficult feelings because avoiding them prolongs the agony, perpetuating our distress. Aristotle was one of the first to suggest this idea, and Freud popularized it centuries later. Fast-forward to the present, where words such as "compartmentalize" and "suppress" and "denial" appear on social media, often in a negative light.

There are good reasons that this myth of universal approach persists. Foa and Kozak's work, along with decades of research by others, has found that avoidance can be damaging long term. Studies show that coping with stressful situations by chronically avoiding, denying, or downplaying can make the problem more upsetting and is strongly linked with anxiety and depression. This finding holds up among many different demographics, from college students to cardiac patients. One study that followed adults over ten years showed that "avoidance coping" not only predicted future depression but also created more problems for individuals over time.

Simply put, there's no doubt that chronic avoidance is bad. It's a blunt instrument being used to fix a wide variety of emotional difficulties—as if the only tool you owned for home improvement projects were a hammer. But what we tend to forget is that hammers *are* good for certain jobs. Not every job, maybe not even most jobs, but *some* jobs. This is where the world of psychology falls prey to the human bias to oversimplify, sorting things into good versus bad categories. The regulation tools that are based on approaching our emotions are often considered *universally* good, and the coping mechanisms we use to avoid our emotions are considered *universally* bad. In recent years, research uncovering more about how the body and brain work to help us regulate emotions indicates that turning away from a distressing situation is a core feature of an evolutionary gift that we all have—the psychological immune system.

The Psychological Immune System

Distraction takes advantage of one of our built-in emotional regulators: the psychological immune system. Just as our regular immune system deals with physical threats, our psychological immune system deals with psychological threats. There are different components to this immune system, but one of the most important is *time*. Our emotions follow a natural time course: The further you get away from the inciting incident of distress, the more the sharpness of the emotion fades. While certain traumatic experiences linger, most experiences have an emotional peak that slopes downward over time, allowing us to gain distance from it.

The trick is, we need to distract ourselves for long enough to let time do its work. Instead of firing back a frustrated response to a work email, you slam your laptop shut; when you come back to it the next morning, you find you're much less angry. In the middle of an escalating fight with a sibling you decide to take a time-out and walk away; when you come back to the conversation, your resentment has been metabolized away. You experience a painful breakup and want nothing but to stay home and wallow in your misery; you're surprised to find that when your friends drag you out dancing for the evening, you have a wonderful time. The old adage "time heals all wounds" may rankle us when we're in the moment grappling with tough feelings, but the fundamental truth of it holds up under scientific scrutiny (well, except for the word "all"; it's more scientifically precise to say "most").

Every person's psychological immune system is different and responds in complex ways to different situations. The same goes for the individualized nature of our emotional experiences. This is one of the problems with a blanket disavowal of any emotion regulation tool. Chronic avoidance is bad, but chronic approach can be problematic as well. Talking about your experience, working through your emotions with other people, and getting professional help are all considered the status quo regimen for how to cope in a healthy way. Much of the time, that is exactly what you need.

But not all of the time.

Studies show that *on average,* if you are a person who tends toward avoidance after a stressful or traumatic event you tend to do worse. However, these studies don't focus on people, like my grandmother and Dennis Rodman, who use avoidance flexibly—meaning strategically and instrumentally—and pivot to expression when necessary. They have a hammer *and* a screwdriver in their toolbox. They are the ones who pivot intuitively between avoiding and expressing their emotions, depending on the context—to their benefit.

Recent research shows that when we study resilience among a broad swath of people, we see that the tools for "healthy" coping are far more varied than the myth of universal approach would have us believe.

Flex It

When the towers fell on 9/11, George Bonanno, a leading expert on resilience, was living in Manhattan, just a mile north of ground zero. Like most people in the city at that time, he was rocked by the tragedy and wanted to do everything he could to help. George had devoted his life to understanding how people manage suffering and thrive in the face of despair. He knew that when misfortune, trauma, and distress strike, there are strategies that help and strategies that can make things worse. So, instead of volunteering at ground zero or giving blood, he decided to do what only he could: research what might *really* help people manage these difficult emotions.

In the aftermath of the tragedy, much of the mental health assistance provided in New York City was based on the idea that it's better to encourage people to talk through their emotions. In fact, legions of therapists from all over the country were sent in to help first responders, families of victims, and people who were in proximity to the attacks. But even before 9/11, George had been seeing evidence that a singular "approach" strategy doesn't work for everyone. So here's what he did: First, he brought a hundred and one New York City undergraduates into the lab and talked to them about what they had experienced in the days since 9/11. Since it was an experiment, he measured how distressed each student was, which provided George and his team with a baseline for how well the students were dealing with their emotions.

Over the course of the next three months George brought the students back into his lab. It was then that he and his team looked more closely at how they approached *and avoided* their emotions. To do this, they sat the students down in front of a computer and a one-way mirror and told them they would be viewing both positive and negative images on a screen. Knowing they would be judged by an invisible participant in the next room, the students had to do their best to *express* or *suppress* their feelings at different points in time during the experiment. Expression is a form of approach, or confronting your emotions, while suppression is avoidance. How well the "judge" in the next room could tell what emotions they were experiencing (or not) showed how skilled each participant was at tuning their emotions up and down.

Eighteen months later, George's team checked in with the students to see how they were doing and measured their levels of distress once again. They discovered that the participants who performed best on the initial task—those who were able to both express *and* suppress emotion successfully—were best able to cope in the wake of 9/11. What Bonanno found was that a person's capacity to flexibly deploy their attention—both avoiding and approaching—was the best predictor of resilience.

Think of your ability to express and suppress emotion as the equivalent of push-ups and pull-ups. Being able to switch back and forth between these exercises is a good measure of your strength and agility. Likewise, Bonanno's study was testing the ability of the students to toggle between expressing and suppressing their emotions. The people who were best at switching back and forth between the two showed a higher degree of flexibility. They could focus their attention on the hard emotions, and they could also move their spotlight away from the tough stuff to get temporary relief. The positive outcomes for the best "switchers" suggest that avoidance is a key part of flexibility, and that flexibility is a key indicator of resilience. This is exactly what Dennis Rodman and my grandmother were doing: Rodman with his let-loose excursions and my grandmother with her occasional emotional confessions on the way home from the grocery store.

Bonanno's research was just the beginning of a tectonic shift toward embracing a less binary view of what constitutes "healthy" versus "unhealthy" emotion regulation and the acceptance of the idea that this kind of flexibility is linked with better psychological health. Being responsive to a changing world is a prerequisite for adaptation and survival. (This is also why we likely evolved the capacity to experience *both* negative and positive emotions—to help us respond to a world that's always in flux.) We use our attention flexibly all the time to approach or avoid things in the service of our physical health: We avoid the poison berries and approach the blueberries. Why would our emotional lives be any different? Why would we rely *only* on approach to maintain mental health?

We wouldn't and we don't.

Do You Need a Hammer or a Screwdriver?

Understanding whether to avoid or approach a given situation starts by self-assessing. Ask yourself: *Is what I'm doing working? Is it making me feel better about the problem in front of me?*

When avoidance works, it might look like this: You say something regrettable at a party, but instead of worrying about it, you choose to distract yourself by watching funny videos later that evening. If you find that brings you some relief, and the concern doesn't resurface down the road (no sleepless nights like Luisa in chapter 2), that counts as working.

When approach works, it might look like this: You spend time thinking about a mistake you made—let's say you betrayed a friend's confidence—but ultimately you learn something valuable. By reflecting on the betrayal and your feelings of guilt, you now see the fuller picture and realize that the temporary high of sharing gossip wasn't worth making your friend feel bad. In this case, approaching the difficult truth allows you to grow and move beyond the experience.

It's important to know that a strategy doesn't need to be free from negative emotions to work well. Sometimes approaching the things that are bothering us *can* be painful in the moment but helpful over time.

In both cases, if approaching or avoiding is working for you, no need to overthink it. Just keep doing your thing.

What if it's not working?

If it's not working for you, that's a cue to switch to another form of attention deployment (approach → avoid or vice versa) or another tool altogether (we'll talk about plenty in the book).

When approach isn't working, the biggest warning sign is chatter (getting stuck in a negative thought loop), which I wrote about in my first book. This can feel as though you are stuck in a traffic circle with no exit, circling the same situation, thoughts, and feelings in your mind over and over again. In other words, you're focused on the problem, but you're not making any progress toward working through it.

Another unhelpful form of approach is if you keep dredging up negative experiences that aren't actually bothering you. What does that look like? A personal example: When I was twelve, my parents got divorced. There was the usual tension and fighting and acrimony, not to mention the fact that in the early 1990s, I was one of the only "divorce kids" I knew. It wasn't a fun time, to be sure. But the only time I think about it is when my dad brings it up. Which, to my continuing dismay, is quite often. He thinks I still haven't "processed" the divorce and encourages me to open up about it with him. While I appreciate his concern, his well-intentioned requests to talk about it end up doing more harm than good. I am *not* a natural avoider; in fact, there are plenty of aspects of my childhood that I could talk about for days. But I talked and thought about my parents' divorce extensively when it happened. It is not a current source of confusion, or resentment, or unaired grievance. The simple fact is that my psychological immune system cleared whatever painful feelings I had about my parents' divorce a long, long time ago.

If something doesn't keep cropping up, there's a good chance you're *not* harboring some invisible wound that will fester and ruin your life decades down the road. Sometimes, because of the myth of universal approach, this can be hard for people to believe.

Next, let's look at when avoidance isn't working. When we talk about unhealthy forms of avoidance, there are three signs to watch for.

Warning Sign 1: You try *not* to think about a problem, but you keep thinking about it. Maybe you go to the movies to avoid thinking about a fight with your best friend, but every plot point reminds you of your failed communication. Or maybe you *thought* you weren't harboring resentment about your spouse's infidelity, but every time you fight, your anger comes barreling into the conversation. The reason? Sometimes our psychological immune system alone can't fully handle the emotional turmoil. Intrusive, repetitive negative thoughts (chatter) are one of the hallmark indicators that this is the case.

Warning Sign 2: You find yourself chronically relying on substances such as drugs and alcohol, or engaging in compulsive sex, gambling, or eating. These are all well-known distraction methods because the relief they provide is powerful, but usually temporary, and almost always harmful in the long term. These avoidant coping mechanisms are big, bright red flags.

Warning Sign 3: You find yourself constantly seeking reassurance. This behavior *looks* like approach, but it is actually avoidance. A good example of this is Luisa, from chapter 2, who found herself obsessively reading medical articles online trying to find a fix for her daughter's peanut allergy. On the surface, it looks as if she were *approaching* her anxiety by doing something about it: trying to solve the problem of her daughter's allergy. Who can blame her? But if you look at the fact that her emotional turmoil was making it hard for her to sleep, work, and parent, you see that there's more going on. She wasn't actually confronting the fear of her daughter's death and approaching it in a healthy way. She was scouring the internet to avoid addressing that possibility, convinced that if she could only "fix it," then her anxiety and fear would go away. Other common examples of unhealthy reassurance seeking: desperately trying to find evidence that everything will be okay (unnecessary medical tests, repeatedly asking loved ones for their opinion, and so on) or excessively checking that the windows and doors are locked. When we fall into these traps, the actions we take to try to cope with fear or anxiety may *look* like actions we are taking to solve a problem, but often our avoidance of the *root* issue just makes matters worse.

If an avoidance strategy isn't working, and approaching is too painful, that may be a sign that a totally different tool is needed. You may need not just to approach the emotion but to approach it in a whole new way, or you may need to look outward, to the world around you—the people, the environment, the cultural waters—to shape your emotional experience.

In many ways, attention is at the root of all of the strategies we'll cover: Where does your attention go? What fills your consciousness? And what effect does it have? There is no universal rule for using our attention wisely: A capacity for flexibility is what really counts.

Imagine if my grandmother had second-guessed her own coping strategies. What if her friends or family had made her feel guilty about suppressing her painful memories? What if she was told that she couldn't deal with it on her own terms, that she had to follow some preordained path to healing? We'll never know, but I can't help but believe she would have been worse off. She might not have even survived her experience in the woods had she been overcome with emotion to the point where she became paralyzed. Instead, my grandmother stumbled on a way to use her attentional spotlight to regulate her emotions adaptively.

She knew how to wield her attention skillfully, approaching her emotions in a way that honored her past experience while avoiding them in a way that let her keep moving forward.

If that's not an emotional life well lived, I don't know what is.

Next, we will look at another tool that is a close cousin to attention—perspective. If attention determines *what* to look at, perspective influences *how* we look at those things. Changing perspective on a situation when you're struggling can be a highly effective tool, but it's not quite as simple as "looking on the bright side." So, to explore the power of—and the pathway to—a true perspective shift, we're going to go to a whole new vantage point: outer space.

Chapter Five

"Easier F***ing Said Than Done": Perspective Shifters

The American astronaut Jerry Linenger was only a few weeks into his five-month stay on the Soviet Union–built Mir space station, but that was long enough to know that the piercing blare of the alarm system was nothing to panic about. It went off all the time: when the carbon dioxide scrubber went offline and when the electricity failed momentarily.

On the evening of February 24, 1997, Jerry was doing a little après dinner data entry—floating horizontal, toes hooked through a loop on the wall, laptop hanging down from the ceiling—when the klaxon blare of the alarm sounded yet again. He popped in the earplugs he kept stashed nearby, saved his data, and casually barrel-rolled toward the source of the sound to investigate. His bet was on the carbon dioxide scrubber; they had been failing on and off throughout the mission.

There are plenty of things that you don't want to let loose

on a space station: bread (the crumbs can gunk up equipment and be accidentally inhaled by the astronauts), guns (an accidental hole in the hull would suck all the oxygen out into space), and human waste (no explanation needed). But one of the deadliest by far is fire. So, when Jerry rounded the corner and saw smoke billowing and flames sparking three feet high out of an oxygen canister, he thought, *Not good.*

That might have been the biggest understatement of his life.

Fire is always dangerous, but in a cramped space station it is fatal much more quickly. When Jerry looked closely, he could see that the metal near the flaming oxygen machine was already melting. If they didn't stop it soon, the fire would melt through the scant inches of metal that separated the crew from the void. All the breathable air would be sucked out of the station in an instant, giving them about fifteen seconds before they passed out. There were a total of five cosmonauts on Mir that day plus Jerry, which led Jerry to later experience another stomach-dropping realization: The location of the fire meant that one of the two evacuation capsules was unreachable. If worse came to worse, only three men could make it off Mir.

Because Mir was so small, it immediately began to fill with smoke. Before anything else, Jerry had to find his way to an oxygen respirator as quickly as possible. He groped along the walls trying not to cough or inhale. Through the smoke, he spotted his fellow cosmonauts doing the same. As his chest tightened and the urge to breathe grew desperate, a voice in his head told him to look to the ground for a pocket of fresh air. But when he looked down, he saw the same thickness of smoke because, as he well knew, smoke doesn't rise in space. That's when another absurd thought popped into his head: *Open a window!*

That almost made him laugh despite his circumstances.

By the time his hand found an oxygen mask clamped to a bulkhead, he had been holding his breath for what seemed like an eternity.

He strapped on the mask, flipped the switch, and waited for the cool relief of oxygen. But all he felt was the mask sucking against his face as he desperately tried to inhale; the respirator was malfunctioning. He pulled it off and went in search of another. As he fumbled to the next bulkhead, his mind was calmly informing him that there was a very good chance he was going to die. During the seconds it took him to get to the next respirator, he had already said his goodbyes to his wife and son, talking to them in his mind, assuring them that he did what he could to get back and apologizing for not being there for them during this trip.

His fingers found another mask and fumbled to release its oxygen. Just as the edges of his vision started to go black, the precious oxygen hit.

His lungs burned, at the point of hyperventilating, as he sucked down the fresh air. He tried to regulate his breathing as quickly as possible. He was alive, but by no means had he solved the problem. Now he had to maintain his composure and think clearly so he could figure out how to fight what was quickly becoming one of the worst fires in spaceflight history.

Okay, you're alive, he said to himself. *Now, you have to do everything right, or everyone on this station is going to die. No mistakes.*

Jerry's years of training and experience had propelled him through the initial moments of crisis, without having to stop and think. His initial response flowed from the comprehensive drills he'd run countless times to be ready for these kinds of emergencies, and as an astronaut he was of course used to performing in high-stress situations. But even for him, this had now become a next-level scenario—the kind of thing most astronauts will likely never have to face, even over the course of a high-risk career—and the rapid-fire decisions he was about to make were quite literally a matter of life or death. He'd need to rely not only on his training and knowledge but also on his capacity to *rapidly* shift through any emotional reactions that might impede his ability to survive.

———

In the last chapter, we talked about using *attention* to shift—deploying our attention flexibly and changing whether we approach or avoid, depending on the context and demands of the moment. In this situation, that's obviously not a strategy Jerry could use. Confronting the circumstance is the only way to survive. And to connect Jerry's experience on the space station to the rest of us mortals down on earth, there are plenty of instances in any of our lives where the same is true: Whatever's going on is big, and intense, and it's happening *right now,* and using attentional techniques would be as pointless, and catastrophic, as turning your back on a fire on a space station.

When we can't turn our attention away from an overwhelming emotional hot spot, we need a different strategy. The good news? We've evolved another type of shifter to deal with exactly these kinds of situations. The bad news: How to actually use it isn't quite as obvious as our "silver linings" catchphrases would have you believe.

It's as Easy as A-B-C (or So They Say)

"Who knows their ABCs?"

That was the question David Williams, a bespectacled University of Pennsylvania psychology professor posed to my class on a warm summer evening in 1999. When he paused for effect, my classmates smirked and rolled their eyes, while I wondered where this was going. Little did I know that the concept he was referring to would stay with me for the rest of my life.

It wasn't the *Sesame Street* version of the ABCs, but a simple formula describing how changing how you think about something can change your emotions. Here's how it works:

A = Adverse event (for example, scary health diagnosis)

B = Belief (for instance, "It's over. My kids won't have a father. I'm deserting them.")

C = Consequence (for example, anxiety, stomach-churning nausea, sadness)

The most important thing about the ABCs is that if you can change B, then you change C, because, not surprisingly, negative thoughts often drive negative emotions. So, while you can't do anything about a scary health diagnosis, you can modify your thoughts about it (for example, "Maybe it's a false positive, and even if it's true, I can still get treated"), which according to the formula should, in turn, allow you to handle the situation better.

This approach is widely known as reframing, but it actually goes by several other names, depending on the academic circle you travel in (reappraisal, reconstrual, and cognitive change, to name just a few), and is the foundation of many therapy protocols today, including cognitive behavioral therapy, one of the most widely supported therapeutic modalities in the world. Think about reframing as a filter that you put on a camera (okay, fine, your smartphone!). You can point your camera (your attention) at the same object, but the filter you apply changes the way the object looks. You might have no choice but to take a picture of a painful scene, but the filter can soften the harsh edges. For example, someone breaks up with you unexpectedly: it's fresh, it's painful, it's shocking, and you can't just look away from the loss. But you can change the filter. Instead of seeing proof that you are unlovable, that you will never have a partner and are doomed to solitude, you can see that you avoided wasting your time and have a fresh opportunity to find your soulmate.

Change the filter, change the way you feel about what's unfolding.

The same network of cognitive control brain regions we looked at in chapter 2, which allow us to transform information, is employed for reframing as well. It includes regions of the prefrontal cortex, which allows us to manipulate information in our minds. In the same way that we can look at a cardboard box and come up with dozens of uses for it, we can look at our current life crisis and think of a dozen different ways to make meaning of it. We can, of course, also change the filter on things that *have already happened in the past* (reframing a past failure into a learning experience) or on things that *could happen in the future* (reframing a worry into a source of possibility or excitement).

Having a changeable filter for how we view our emotions is important for two reasons. The first is that there are experiences we *must* look at; we simply can't look away even if we'd like to. For example, a parent who is declining. A chronic illness you're facing. A fire, blazing a hole through the thin wall that separates you from the void of space.

Second, even if we can pull our attention away from what's driving our negativity, avoidance doesn't always work. Avoidance can be healthy as we previously discussed, but at times it can act like a Band-Aid that obscures the wound without healing it as well. Sometimes, to make it through the emotional firestorm, we instead have to draw on our ability to stare right at something horrifying and change how we think about it— another capacity that distinguishes us from every other species on the planet. This ability to reframe how we think about challenges, defeats, and losses is how we survive, how we succeed, and how we thrive.

As an impressionable nineteen-year-old undergraduate, learning about this simple formula felt like a revelation. It was so . . . logical. You can't always change the adverse events you experience—the A—but you can *always* change how you think about it to change how you feel.

Or so I was told.

The Reframing Paradox

The trouble with reframing was eloquently summed up by a friend of mine when having a debate with his wife about a tricky work situation. I happened to be in the car as he talked, listening as he told the story. At one point his wife suggested he just look at things *more positively,* to which he half jokingly replied, "Easier f***ing said than done!"

When it comes to reframing, one of the big challenges is that many people don't know how to reframe their experiences adaptively and fall into the trap of reframing *negatively.* A flat tire at an inopportune moment becomes more evidence of lifelong bad luck instead of a challenge that happens to everyone. A disagreement with a colleague is seen as a personal affront as opposed to a misunderstanding. A fire in space is a clear sign that the mission is doomed and getting on the shuttle was a catastrophic mistake.

You can see evidence for reframing gone wrong very clearly among people who tend to worry or ruminate about their problems. When you're worrying, you're using your cognitive control system to try to solve a problem, but you end up making things worse by negatively reframing your circumstances. For instance, if I'm scared of attending a party, I might run a series of simulations in my head about what would happen if my worst fears came true. *I walk in, and I don't know a single person there; everyone looks at me as if I were crashing the party. A friend I'm talking to goes off to talk to someone else, and I'm left standing awkwardly by the drinks table all alone. Everyone stares at me.* This kind of negative reframing can get you stuck in a worry loop. Instead of solving the problem, my attempts to reframe it just made me more agitated and fearful.

Several years ago, I investigated how reframing goes awry in people prone to worrying. Friends and therapists often tell people who worry to "focus on the bright side," but they obviously don't. The question was why? One possibility was that worriers don't even try to find the silver lining. They turn instead to thinking about all the anxiety-provoking what-ifs that could happen. But the other possibility that we were interested in exploring was that worriers might try to do exactly what they're advised to do—positively reframe their circumstances—but simply fail.

To shed some light on the question, Jason Moser, a clinical neuroscientist at Michigan State University, and I performed one of the first neuroimaging studies to peer into the brains of worriers while they were trying to change the way they think about their emotions. We asked seventy-one women to positively reframe their negative feelings (women were the focus because they are twice as likely to suffer from anxiety). We tried to make it as easy as possible for them to do the task, showing them pictures of negative images, but ones that weren't personally relevant and thus less activating than images depicting their greatest fears. Then we asked them to try to see the images in a more positive light.

While they were performing the task, we monitored their brain activity using an electroencephalogram—a device that looks like a swim cap with electrodes attached—which provides a window into how quickly different psychological processes kick into gear in the brain. We found that the more prone participants were to worrying, the more difficulty they experienced positively reframing how they felt. They exerted much more effort and were also considerably less effective at reducing their negative feelings according to the brain readouts. The study provided neural evidence to support my friend's "easier f***ing said than done" thesis: Even when you ask someone who is prone to worry to positively reframe their feelings, and you make it really, really easy for them to do so, they still struggle.

To reframe a situation, we often need to look at it from a different vantage point so we can shift our thoughts about it. The problem is, when we are washed over by negative feelings, we tend to narrowly focus on the problem at hand, which makes sense because it's what we do throughout our lives when we have a problem. You can even see evidence of this zooming in by looking at the type of language we use when we're immersed in our negative feelings. A kind of linguistic marker of how consumed you are with yourself, the overuse of the words "I," "me," and "my" is the equivalent of zooming in when taking a selfie. There's nothing wrong with using these parts of speech in general, but research increasingly shows that they can signal how absorbed we are with our problems.

In one study, researchers analyzed more than a million social media posts of people who wrote about their breakups on Reddit. They combed through all the posts users made during the years previous and after their romantic fallouts and found that they could predict with precision when a person was going to experience a breakup based on their increasing use of first-person singular pronouns in their writing. They also found that the longer the users stayed in these zoomed-in states, continuing to post about their breakups in the first-person, the worse off they were a year later. Another, complementary study by a different group of scientists found that they could predict a person's likelihood of being diagnosed with depression based in part on the amount of first-person singular pronouns contained in their Facebook posts!

Linguistic markers of immersion and their associated problems show that looking at painful experiences through this zoomed-in selfie lens magnifies the problem. The closer the camera gets to that giant zit on your face, the more stressed you get and the more horrible it seems. And the more intense our negative emotions are, the more challenging it is to reframe them because the stress that comes with big emotion saps us of the neural resources we need to implement many strategies that might help.

Here's a simple model of how it works: A network of brain regions in your prefrontal and parietal cortices, which are roughly located behind your eyes and above your ears, help you reframe things. You have a goal (see the problem from a different viewpoint), and then you have to use your working memory to keep that goal active in your mind, inhibiting other distracting information from intruding into your awareness and generating alternative ways of thinking about your problems. Anyone who has ever rehearsed a conversation in their head knows this takes focus. The problem is that your prefrontal cortex is negatively affected by stress, which degrades the connections between synapses in that part of the brain. Even a little bit of temporary stress can trigger what researchers describe as a "dramatic loss of prefrontal cognitive abilities."

Even Jerry, who was exquisitely trained to manage crises, experienced little blips in his reasoning, likely due to stress. His brain was surfacing "solutions" to the problem, like opening a window to let out the smoke or getting low to the ground, that would have worked on earth but were nonsensical in his current situation. The more intense the stress, and the longer its duration, the more zapped our prefrontal cortex becomes and the less likely we are to pull ourselves out of an emotional spin cycle and frame things effectively. Think of it like a brownout on the electrical grid. Maybe your lights are a little dimmer and the Wi-Fi keeps cutting out. It's a similar concept in the brain. Things are still working; they're just not working optimally.

So what's the solution to this reframing paradox?

Lessons Learned from Swimming with Small Children

When my daughters were little and still floating around in life jackets in the pool, I would swim with them in deep water. Occasionally, something would scare them, and they would reach for me, scrambling to board me as they would a rescue vessel. Unfortunately, I was over my head (literally!), which is not a good time to give panicky shoulder rides. I had to keep them at arm's length—hold on to their life jackets and talk to them in a soothing voice. This worked to keep them calm. But it also provided me the *distance* to deal with their fear without drowning myself.

Swimming with small children who are struggling in the water is a lot like being in the thick of negative emotions: We're desperately seeking safety but often end up making things worse when we try to solve the problem by getting closer to it, rehashing it over and over. Pulling back the frame, gaining distance, allows us to engage with our problems without becoming consumed by them.

Early in my career, I thought there were only a few ways to gain distance. Pretend you're a fly on the wall observing your problem. Meditate. Travel if you can. But what I have learned since is that there are many other ways to distance. For example, if you're one of the millions of people who speak multiple languages, you've got an advantage sitting right on the tip of your tongue.

Research shows that a link exists between our native language and emotion. It's the language that we first learn and use to think about the world, the language through which we first experience our greatest triumphs and most unsavory defeats. As a result, feelings are more potent in our native tongue than in second languages we acquire. Curse words have more punch. Taboos are more disconcerting. Embarrassing events elicit more cringe. Because we learn about emotion in our native tongue, the links between the experience of emotion and the native language words we use to reference those experiences are very strong.

On the flip side, when we speak in a second language, we are less affected by the emotional weight of words, making it easier to keep a cool head. Studies show that thinking in a foreign language leads people to reason more objectively and be less biased in decision making—a phenomenon dubbed the foreign language effect. So, if you're bilingual (or more), you've got an emotion regulation app already built into your brain that you can access by toggling to your *second* language. And if you're not, there's a larger lesson here about language and emotions. These findings speak to the potential that small shifts in language have to reroute how we relate to ourselves and manage our emotions as a result. If you don't speak a second language, you don't need to shell out for Rosetta Stone; there's an even more accessible way we can harness language to shift your emotions.

It's All About "You"

Early in the 2022 Wimbledon men's quarterfinal, the No. 1 seed, Novak Djokovic, was getting unexpectedly routed by the twenty-year-old Jannik Sinner. Down two sets (5–7, 2–6), Djokovic took the tennis equivalent of a time-out: He asked for a toilet break. Moments later, he trotted out of the locker room, resumed play, and proceeded to rout his opponent. He dominated in the next three sets (6–3, 6–2, 6–2), besting Sinner, and ultimately went on to win Wimbledon.

Later, when asked about the turnaround and what he was doing in the bathroom, Djokovic unapologetically replied that he was giving himself a much-needed pep talk. In the locker room he stood in front of the mirror, looked himself in the eyes, and said, "*You* can do it. Believe in *yourself.* Now is the time, forget everything that has happened. New match starts now. Let's go, *champ.*" Cheesy as it sounds, it worked.

To me, the most remarkable thing about this story isn't the incredible comeback. It's not even the fact that Djokovic gave himself a pep talk. It's the pronoun usage. He doesn't say, "I can do it." He says, "*You* can do it."

This small, seemingly insignificant change in language is consequential. Why? As you've no doubt experienced, and as I wrote about in *Chatter,* it's much easier for us to give advice to other people than it is to give ourselves advice. The name for this phenomenon is Solomon's paradox, eponymously named after the Old Testament's King Solomon, who was famously adept at doling out wisdom to others but often stumbled when it came to his own life (he got involved in several extramarital affairs that ultimately led to his downfall).

Using the word "you" to silently refer to yourself is called distanced self-talk, and it works like this:

"You" is a word we almost exclusively use to think about and refer to other people. So, when you use that word to refer to yourself, as Djokovic did, it gives you distance and shifts your perspective. It gets you to think about yourself the way you would think about *someone else.* This slightly weird, seemingly tiny linguistic shift is consequential because the difference between "I am stressed out" and "You are stressed out" is big. If I'm stressed out, I might feel panic, a racing heart, and a looping anxiety. If someone else is stressed out, I might feel compassion, empathy, and a desire to soothe their nerves. By talking to myself using the word "you," I am casting myself in the role of "someone else." I'm able to see and to *feel* my situation from a different perspective. Using your own name (instead of "I") or even the third person "he" or "she" also works as a linguistic shifter in much the same way that "you" does. *Come on, Ethan, you can finish writing this chapter!*

Jerry Linenger, up on Mir, almost immediately pivoted to distanced self-talk—probably without even knowing what it was. After that initial spike of alarm, he "tucked" his fear away into a separate compartment of his mind and started to give himself a pep talk.

"Okay, *Jerry. You've* got to get going. *You* need oxygen here; *you* need to start acting."

While three members of the Mir crew split off to ready the space station emergency evacuation capsule, Jerry and one other cosmonaut stayed to fight the fire with extinguisher after extinguisher. Unable to see each other very well, or talk, they had to communicate by shaking one another every so often to make sure they were okay. The flames they fought eventually reached three feet high. After fourteen minutes, the fire burned itself out from lack of fuel. It would go down in history as one of the worst blazes aboard a spacecraft. Miraculously, everyone aboard survived.

Later, Linenger would recall his mind being on two tracks throughout the emergency. One track he calls "rational survivor mode," which he had honed through years of training as a U.S. Navy flight surgeon, and before that as a doctor who had to tend to triple-gunshot wounds in the ER. In this mode, his mind had a checklist of action items ready for him and both the confidence and the determination to knock everything off the list.

The other track was the wild card: Anything could pop up in his mind unbidden and often did—fear, worry, unhelpful instructions, and future projections. This line of thinking was unpredictable, reflexive, and sometimes straight-up baffling. (Open a window in space? Really?!)

Distanced self-talk has this effect of almost splitting "you" into two: the part of you that's having the emotions, and the part that's coaching you through them. When Jerry talked about the "two tracks," I immediately thought of the way I shift into self-talk during moments of stress. I haven't shut off my feelings, but I can step out of their stream, see them more clearly for what they are, and talk myself step-by-step through what I need to do. The best part: This silent internal practice of becoming your own emotional coach is one of the *least effortful* tools for shifting your perspective.

In one pair of neuroscience experiments I performed with Jason Moser and his team, we found, for example, that participants showed signs of experiencing less negative emotion within seconds of using distanced self-talk to regulate their emotions. Moreover, using this technique didn't activate a brain waveform that tracks the exertion of effort, which is consistent with the idea that this tool isn't overly costly in terms of exhausting your precious prefrontal resources.

Other research has shown that the more people use distanced language when trying to reframe their feelings, the more effective they are at doing so, and has documented benefits of this tool across a variety of situations: when people are reflecting on past heartache or future worries and when they grapple with negative feelings in the heat of the moment after a date or big interview. Across these contexts, you can see benefits in terms of how people feel, how their body reacts physiologically (they're calmer), and how they talk to themselves (they're more challenge oriented and optimistic).

One study explored how a variant of this tool plays out in the context of therapy. The Princeton psychologist Erik Nook and his colleagues at Harvard examined more than 1.2 million therapy transcripts from Talkspace, an app that virtually connects people struggling with mental health problems to therapists who specialize in cognitive behavioral therapy, an intervention that focuses on helping people reframe. They found that as participants' therapy sessions progressed, they increasingly used distanced language when talking about their feelings, which was associated with improvements in how they felt.

Distanced self-talk has been shown to have other benefits. One study that tracked people over time showed that it improved how wisely people reasoned about social conflict. Over the course of a month researchers instructed participants to keep a diary using either first-person language or distanced language when writing about their problems. When the researchers collected the diaries, they found that people in the distanced language group showed more evidence of "wise reasoning"; they were more open-minded, took others' perspectives, and showed more signs of intellectual humility.

People often ask me what I do when I get caught in a negative emotion loop. While always pausing to make scholarly caveats about how no one tool is best, eventually I confess that distanced self-talk is my go-to because of its ease and effectiveness. But distanced self-talk is not the only science-based distancing tool I lean on frequently. Another of my favorites is one that flies in the face of popular wisdom. Instead of remaining "in the moment" it requires that we transport ourselves past it—as fast as we can.

The Value of *Not* Being in the Moment

One of the features of the human mind that I find most remarkable is its ability to mentally time travel. We do it all the time when we recall a vivid memory from childhood or imagine what it might be like to go on a wonderful vacation. Sometimes when we jump into this time travel machine to make sense of negative past experiences, it breaks down, so we end up ruminating about them. Sometimes we get stuck in the future, swimming in a sea of anxiety as we simulate all the potential worst-case scenarios that might befall us.

When these things happen, one solution is to refocus our attention on the present. This is the basis of mindfulness practice, a time-tested tool for emotion regulation with ample scientific support. There's no doubt that it can be a useful tool to deploy. But we lose sight of two crucial truths when we dogmatically prescribe the present as the antidote to all our emotional turmoil.

First, those who strive to always be in the moment will be frustrated because it's simply not possible. The human mind evolved to time travel. There are species that live the bulk of their lives in the moment, navigating from one here-and-now experience to the next. Those species tend to have multiple legs and large antennae (think: cockroaches, spiders, and other creepy-crawlers). The degree to which we can time travel in our minds sets us apart from these creepy-crawlers and every other species on this planet. It's a vital tool that allows us to make sense of our past experiences, plan for the future, innovate, and create.

The second truth we lose sight of when we're trying to be in the moment all the time: We can actually learn to time travel more effectively. Part of the reason we are admonished to stay in the moment is that we keep getting stuck in the negative past or the future, which is admittedly unhealthy. But if you can't engage with the past or the future, you can't make good decisions based on past knowledge or plan for the coming years. Being in the now is great sometimes, but not *all* the time.

The next time you find yourself in a moment of crisis, try leaning on your capacity for time travel. It can be our Achilles' heel; it can also be a kind of human superpower.

Fall back into the past.

What other times in your life have you gone through that are similar to this one? How did you get through them?

What have you gained from past experiences that you can use now?

If you've gone through worse, you know you can get through this.

If you haven't gone through anything worse than what you are currently confronting, some of your past experiences will have prepared you, in one way or another. And people in your life who have shared stories from the past from their own lives can be pivotal guides, offering a broader perspective that places your own struggle in a longer trajectory of living, surviving, persevering. For instance, when I find myself struggling, I think about my grandmother, and how if she got through what she did, I can manage my own problems.

Project yourself into the future.

> How will you feel about this event in a week? In a month? In a year?

> Go all the way to the end of your life, perhaps decades from now: What meaning will this event have for you then?

Luisa, who saved her daughter's life on that airplane, also faced the usual struggles of new parents. Up all night cluster feeding her newborn, exhausted, lonely, "zonked and miserable" (as she puts it), she looked for a way to pivot out of that low feeling. A part of her brain was screaming at her that these days were fleeting, that she had to find a way to enjoy some of the time with her new baby.

"I remember I started imagining her as a kindergartner, to get through those long nights," Luisa said. "It really helped me shift from being in despair that I was awake at 3:00 a.m., again, barely able to keep my eyes open, to feeling tender about how brief this time was actually going to be in the grand scheme of things." Picturing her daughter tall and long-legged, not a baby anymore, gave her this visceral kick of missing that baby . . . and then looking down to find her baby still there, curled in her arms. A rush of gratitude for the present moment, born from a little time travel out of it.

Using mental time travel as a tool means being deliberate about bringing the concept of impermanence into awareness; when you think about how you'll feel about a stressor some time down the road, you realize that what you're going through, as bad as it may feel in the moment, will eventually pass, giving you the boost you need to deal with the present. When we're sad or grieving, it can be difficult to imagine that we won't feel that way forever. But the truth is that things are always changing. Desires, circumstances, beliefs. When we can zoom out and remember the wisdom of "this too shall pass," we invariably feel better because we are reminding ourselves of the impermanence of life.

From Earth to Outer Space . . . and Back Again

Changing your perspective offers you the possibility of *downshifting* your emotional response. In situations of extreme distress—for instance, the loss of a child or a terminal diagnosis—shifting perspective is not easy and can sometimes feel impossible. But it *is* possible, though it can take time and perhaps an unexpected form.

When Vic Strecher lost his nineteen-year-old daughter to a heart issue, he plummeted into the darkest depression of his life. One morning, three months after her death, he woke up at 5:00, walked down to the edge of Lake Michigan on Michigan's shore, got into his kayak, and started paddling across the freezing water. He paddled for what seemed like miles and recalled thinking, *Maybe I'll go all the way to Wisconsin.* There was a part of him that didn't care if he ended up at the bottom of the lake. And then the sun broke over the horizon, and he heard a voice, talking to him. It sounded like his daughter Julia.

"Get over it, Dad," she said.

It startled him out of his reverie. He saw himself at a distance, through her eyes: in his underwear, in a kayak in the middle of Lake Michigan, focusing far too much on himself, his own pain, his own loss, his own ego.

He didn't sink to the bottom of Lake Michigan that day. He paddled back to shore, sat down at his kitchen table, and said to himself, "Look, you're a behavioral scientist. You should be able to fix yourself. If you can't, what are you doing?" And after that dose of distanced self-talk he got up, got dressed, and got back to the business of living. Since then, he's dedicated his life to helping others find what he regained that morning on the lake: purpose. That brief, powerful moment of perspective shifting, which very well might have saved his life that day, also set him on a completely new path. It didn't change the fact that he lost his beloved daughter. But it placed Vic on a new trajectory that added purpose to a life lived in the wake of loss.

There's an important point I'd like to make here, and it has to do with how we use the shifters we've talked about so far, including perspective shifting. We may find ourselves using these shifters to cope with the usual trials of life: disappointments, worries, mistakes. But these shifters are also tools we can lean on to grapple with the worst times we struggle through as well and not only *survive* them but also make meaning from them.

Paul Kalanithi, in his indelible end-of-life memoir, *When Breath Becomes Air,* described the crushing grief and resentment he felt when he thought about the plans for his life that would never be fulfilled—the things he wanted to do with his family, with his medical career, with his life. As a surgeon and a curious lifelong student, he also thought, "Shouldn't terminal illness, then, be the perfect gift to that young man who had wanted to understand death?" The book paints a portrait not of a linear or neatly resolved emotional process but of a shifting back and forth in time and focusing on himself from a distanced point of view (note the way he references himself not in the first person but as a "young man"), like a song hitting both major and minor notes, until the score runs out.

I don't pretend to know what it's like to face a terminal diagnosis, but I've been through losses, and I know that there are some things in life that shake us to our core. The work of *shifting,* then, becomes part of the longer project of having a life that is, fundamentally, defined by meaning, even through the harsher experiences of a lifetime. These strategies for perspective shifting have the potential to have a cumulative effect on our lives. Seemingly minor micro-shifts can combine to form days, weeks, and years where we have more fluidity across emotional states, less stress, less "stuckness," more joy. And when it comes to the major trials—the losses and traumas that are inevitably part of a human life, but are not something we can just "shift" out of at a moment's notice—tools like perspective shifting become lifeboats, where we can rest for a while, recover some strength, before diving back in.

There are losses in life that change us forever. *Changing the filter* doesn't mean denying the pain we feel; what it means is making sense of it, as part of the longer story of our lives, and having some agency over how that story unfolds. Viktor Frankl, Holocaust survivor and psychiatrist, wrote about his experience reframing some of the worst experiences a person can endure as he was tortured physically and emotionally during World War II. As he famously put it, "Everything can be taken from a man but one thing: the last of the human freedoms—to choose one's attitude in any given set of circumstances, to choose one's own way."

———

So far, we have unpacked the emotion regulation tools that are already inside us: harnessing the primal power of sensation, being flexible with our attention, and changing our perspective. These are all critical levers of emotion that you can access right now, sitting wherever you are sitting, feeling however you are feeling. We take these with us wherever we go, and they are always there for us when we need them. That is their beauty.

But no human is a closed system; we live in the world. We imbibe it, process it, and are affected by it. The spaces, people, and cultures we surround ourselves with all affect the intensity, duration, and resonance of our emotional lives. They are shifters in and of themselves, but they also have a superpower: They can *activate* our internal shifters as well. It would be easy to dismiss these factors as largely out of our control. We can't help it if we live in a tiny studio apartment, or if someone else blows up at us, or if our workplace is toxic, right? Well, yes and no.

There are tools *outside* ourselves that are just as accessible as the ones on the inside, and that we can reach for in the moments when—for whatever reason—our internal shifters are not moving as well as we'd like. To see how this works, let's look at why one of the world's foremost experts on happiness couldn't pull herself out of what felt like the world's foremost depression.

Part Three

Shifting from
the Outside In

Chapter Six

Hidden in Plain Sight:
Space Shifters

In the fall of 2019, Laurie Santos was so busy she had to turn down the pope.

At least that's how it felt to someone raised to be a good people-pleasing Catholic girl. The speaking invitation was for a prestigious event at the Vatican, but she couldn't fit it into the complicated game of *Tetris* that her professional calendar had become. For the past year she had been doing exactly what she cautioned her students at Yale *not* to do—taking on too much. She had good reasons. Laurie was on a once-in-a-lifetime professional joyride. The problem was that she couldn't figure out how to get off the ride, or if she even wanted to.

Laurie was a professor of psychology at Yale. She ran her own research lab, taught a full slate of courses, and had recently been appointed head of Silliman College—the largest residential college within the university. Taking over this coveted position required living with her students in a giant mansion-cum-dorm, organizing community events, and looking out for their general welfare—an Ivy League den mother of sorts. She loved the job. But she soon realized how in need of help her students were. Which is why when she conceived of a class called Psychology and the Good Life, she thought it might be popular.

What Laurie did not expect was that the class would become *so* popular it would completely change her life.

On the first day students were allowed to register for classes, more than nine hundred signed up. By the time the enrollment period closed, a full quarter of the Yale student body was on the roster—an unprecedented occurrence in the history of the storied institution. Only a few weeks into the semester, *The New York Times* found out about the class's popularity and ran an article that led to an onslaught of media requests. During most lectures, Laurie had two microphones clipped to her sweater and two battery packs attached to her boots. One was recording the class for an online Coursera offering, while another was providing audio for whatever national news team happened to be filming her. She kept an Excel spreadsheet to track the outfits she wore so she didn't appear in the same one too many times. Multiple times a week she found herself flying out of New York at four in the morning to deliver a talk, only to hitch a return flight later that day so she could be back by dinnertime.

Not long after the first Psychology and the Good Life course, Laurie was approached to launch a podcast and quickly began working with a producer on a pilot episode. All the while she was still teaching, running a lab, and living among the students. Her duties as head of college were all-encompassing. She hosted Halloween haunted houses, fielded emails constantly, and even rescued dying plants when students couldn't make it back to their dorm rooms. Soon, *The Happiness Lab* podcast became a sensation (currently 100 million downloads and counting), vaulting her to happiness-guru status.

Fame and renown had come suddenly, and just as swiftly came the associated emotional challenges. Laurie was living her dream, but she also felt unrelentingly stressed—like a hamster hopping between wheels and continually getting toppled. She was falling asleep at her desk and spending holidays curled up with journal articles to catch up on the work she was missing. Her fuse got shorter and shorter.

One day in a soundproof podcast booth she discovered she'd been pronouncing a guest's name wrong throughout the recording. With no one around, the emotions burst out of her. She screamed at the top of her lungs and accidentally pounded her laptop so hard it broke the keyboard. Sheepishly, she brought it to the IT department for repair.

Laurie felt guilty for not having enough time for the students she loved or for her husband and friends. Her calendar was oppressive, and her people-pleasing nature made everything worse. She knew she needed help, and she knew she needed it fast. So, she hired a therapist and, during twice-weekly sessions, tried to wrap her mind around what was going wrong. She started saying no to things—like the Vatican event—and she desperately tried to remember the wisdom she imparted to both her classes and her podcast audience.

Take time for yourself.

Find the silver lining.

Keep a journal.

The problem was, she just couldn't seem to make herself do any of those things. While it helped to talk in therapy about what she was feeling, she still couldn't change the way she saw her situation. She still felt confused and overwhelmed. None of her attempts to fix things worked, which left Laurie feeling like the world's biggest hypocrite. Here she was, a supposed expert on happiness, and she was completely snowed under with negative feelings.

Sometimes her colleagues on the podcast would see her struggling and remind her of the very lessons she had just recorded. While their intentions were good, their reminders led to only more frustration and self-criticism. Her passion for her work and her commitment to her students were what she prided herself on, but the stress she was experiencing was a giant roadblock to once-accessible parts of herself. She was both devastated and incredulous that with all the resources at her disposal she couldn't find a way to use them.

Then, one day, inbox overflowing, she received a benign request—a student looking for resources to help fund a dental emergency. Ordinarily, she would have been happy to help. This time she felt the irritation and anger rise in her chest. *Great, one more thing I have to do.* Almost as soon as the thought floated through her head, she felt shocked that she had entertained such a callous idea. Here was a student reaching out, asking her to do her job in a caring way, and instead *this* was her reaction?

It was then that she knew she had hit her breaking point. So, she did the last thing anyone expected.

She left.

The Power of Place

In the last part of the book, we looked at the major internal shifters—your sensory shifters, attention shifters, and perspective shifters. And it's true that wherever we are, and whatever the situation, we always have access to those: They are wired into our neural networks, carried inside us like tools in a case.

But those internal shifters of ours aren't just lying there dormant, quiet until we grab them; they are being pushed, pulled, and manipulated by external factors throughout the day. And one of the most profound external forces we encounter, one that presses on those three internal shifters of senses, attention, and perspective, is *place*.

Like players on a stage, we exist *in context*. The spaces we inhabit and move through shape our emotional lives. They impact our emotions via two pathways: *indirectly* through the larger physical environment's influence on the rhythms and parameters of our daily lives, and *directly* through the immediate experience of our surroundings.

Consider one of the most powerful illustrations of the *indirect* pathways that I've come across. In 2007 the University of Chicago behavioral scientist Thomas Talhelm was living and teaching in China. On a school break, as he toured the country, making friends and bumping into strangers in the grocery store, he noticed that people in southern China and people in northern China seemed *vastly* different: People in the North struck him as more outgoing and independent, and less worried about making others around them feel comfortable, while people in the South were wary of strangers and unfailingly polite.

Now, there's perhaps nothing too surprising about regional differences in behavior within countries; I've certainly noted them myself when I travel from the Midwest (friendly!) back to the East Coast (less friendly!). What is notable is *why*.

Talhelm ran a series of studies that documented these regional differences across China and found that they were fundamentally about *landscape*. The Yangtze River that cuts across China, bisecting it into North and South is not only one of the world's longest rivers, it is also an agricultural dividing line. In the North, the land was better for growing wheat. In the South, the land is better for growing rice. Over thousands of years the different environmental and social demands associated with farming rice versus wheat shaped two vastly different cultures.

Despite both being commodities, rice and wheat have very little in common when it comes to farming practices. Rice farming is intensive. It requires robust irrigation and twice the amount of labor that wheat does. This means that to successfully grow rice, you desperately *need other people*. You need them to coordinate flooding and draining of fields, and you need them to share labor or no one's crops will survive. This creates a condition Thomas calls "functional interdependence." People have to work together to be successful.

The North, by contrast, has no such need for intense cooperation. No need to press your neighbors into helping you harvest or check in with your buddy down the road about when he's flooding his field so you can plan when to flood yours. With wheat, you just need to plant, weed, and harvest. As a result, people who grew up in the North have been inculcated into a culture that does *not* rely as much on social relationships for its livelihood.

And the environment also impacts the emotional side of people's lives. People in more socially dependent rice-farming areas are more apt to engage in envy-inducing social comparisons as well.

In short, the landscapes we occupy shape our lives in all kinds of indirect ways that have trickle-down effects into our emotional worlds. But there are also countless ways our surroundings *directly* influence us. Our environment directly pulls on our sensory, attention, and perspective shifters as well. The ubiquity of this influence—we're always in one space or another—on our emotions means that we must consider how the physical spaces we occupy may be either helping us manage our emotions or getting in the way of our peace.

When it comes to using the space around us to manage our emotions, we have two options:

Switch your space.

Modify your space.

And in my experience both techniques are hugely under-appreciated.

Emotional Oases: Switching Your Space

When Sean, a twenty-nine-year-old army commander, came home on leave after a ten-month deployment that saw him dodge roadside bombs and rocket attacks in Iraq in 2004, he had no idea what to expect when he stepped off the plane and onto the tarmac at Fort Drum, New York. But it wasn't this: his wife, somber-faced and cold, handing him his three-year-old daughter and five-year-old son, turning around without saying a word, and walking away.

Four months into his deployment, when she stopped answering phone calls and started sending brief email dispatches, he began to suspect that something was wrong. But it wasn't until another four months later, when a friend in his unit came back to Baghdad from leave and told Sean he'd best get back to New York, that he really started to worry.

Sean and his wife had been high school sweethearts, and when he deployed, everything was fine. On the tarmac, he stood blinking in disbelief, almost numb with confusion. All he could do was keep on squeezing his daughter to his chest and hold his son's hand, hoping they wouldn't pick up on the gravity of the moment.

At first, he didn't really understand what was happening. Was she just frustrated and in need of a short break from the kids? But after he hitched a ride home from the airport with his kids and walked into the house, the weight of understanding hit him straight in the gut. She had left. Her things were gone. His belongings were in disarray, and here he was with two kids and no idea what to do next.

Over the next few days, the circumstances would become clearer and more painful: She had started a relationship with someone else on the base and wanted nothing more to do with him. As if that weren't painful enough, the man she left Sean for was another soldier he knew.

In ten days, Sean was supposed to go back to Iraq, but his life was in chaos. He had no one to watch the kids. His car had been left to rot in the driveway and wouldn't start. His wife had thrown out most of his clothes. He couldn't understand what he'd done to deserve this. Sean was filled with rage at his wife and her lover and panicked about what this would do to his kids and his career. At the same time, he was grieving the loss of the woman he had vowed to be with forever.

Sean managed to get a leave extension from the army, but still, he didn't know what to do. Waves of sadness and fear would crescendo and turn into despair, which was when the dark thoughts surfaced. Never in his life having dealt with any kind of emotional storm resembling this, he at times, briefly but consistently, considered that everyone might be better off without him around. It was at one of these points, his lowest ever, that his mom called.

"Sean, pack up the kids, get on a plane. Come home."

So, he did.

They flew from Fort Drum to Denver, then rented a car and drove to Casper, Wyoming. The minute he walked into his childhood home, Sean was overcome with a blend of feelings—nostalgia, loss, safety, and relief. So much had happened, he was in so much pain, and now he was here, a place that reminded him of happier times, which over the past week and a half had felt like ancient history. After his mom gave him a hug and set the kids up with a snack, Sean peeled off to take a short nap. He crawled into his old twin-sized bed and slept for more than fifteen hours. It was almost as if after ten days of being clenched by emotional pain, his body was able to relax. When he finally woke up, he looked around his old bedroom filled with outdated movie posters and Little League trophies and realized that he felt safe for the first time in longer than he could remember.

Twenty years later, Sean looks back on the five days he spent in Wyoming as a turning point in his life. It was a necessary sanctuary, allowing him to find the psychological safety needed to find a path forward. When he woke up from his epic nap, his problems hadn't washed away—far from it. But he was able to look at them from a different perspective and see the positives. This would be a clean break from his marriage, he knew where she stood, and there was some mercy in that. He had the leave from the army that he needed to take care of business—sign divorce papers, draw up a custody arrangement, and get back to work.

From there, things slowly got back on track. Sean would eventually go back to school and remarry; he ultimately earned his PhD and is now a professor. But it all started with a place known for expertly plucking the most powerful of emotional strings: *home*.

When we think of the physical environment's direct effect on our emotions, we tend to think about landscape: Are we in an urban or rural place? Are we hearing traffic or waves? Is there green space nearby? I discussed the transformative power of greenery on our cognition and emotion in *Chatter*, and it remains a powerful lever. Many of us are already aware of the abundant findings that speak to the power of green spaces to restore our attention, lift our mood, and induce a sense of awe. But there's another important, lesser-known factor to consider when we think about emotions and environment, and it's one that Sean used to life-changing effect—without even fully knowing it. And that is our personal *attachments* to place.

There's even a name for this phenomenon: place attachment. And it's not about space having a particular quality per se; it's about how it resonates with *you* emotionally. A quick poll of my friends revealed that "their" places include a secluded beach on the coast of Maine, the Sleeping Bear sand dunes in northern Michigan, and a coffee shop overlooking the beach in Los Angeles. Call it a happy place, a spiritual home, or anything you like; it's somewhere in the world that evokes strong feelings of contentment, well-being, and meaning.

Our earliest attachments to our caregivers shape our emotional lives, affecting how trusting we are, how avoidant, or how open. While place attachment isn't as popular a subject as interpersonal attachment (if you don't believe me, just watch Google Scholar explode when you type in "attachment theory"), it's worth knowing more about it. These special places invoke emotions like love and feelings of responsibility that stir people to action. For example, people whose special places are in nature are more likely to act in "pro-environmental" ways, picking up trash, calling for conservation efforts, and even becoming climate justice advocates.

But our relationships to places can also be complicated. Take your hometown, for example. Some people feel a sense of homecoming, peace, and safety when they go back and visit where they grew up. But not everyone has a fully positive relationship to their childhood home. Many people feel as if they regress when they go back home, but other places in our hometowns can be even more straightforwardly negative. In the case of people recovering from substance abuse problems, old hangouts like bars and friends' houses can bring up negative emotions because of their past associations with bad experiences.

Positive place attachment is not just about your childhood home or a beloved college town, and it doesn't have to be far away. You don't have to pack up a U-Haul or hop on a plane to fly home to experience its benefits. You might not need to leave your neighborhood—or even your house. I think of my kids; when they're upset or anxious, they want to retreat into their rooms. Hyper-local versions of our special places can almost always be leveraged. Maybe it's a special reading nook in your attic, or the giant oak tree in your backyard; what matters is that it has a restorative effect.

If we turn back to Laurie Santos, we see that place shifting was the key to shifting out of the emotional overwhelm that was breaking her down, mentally and physically. The day that Laurie finally hit the wall, she decided there was no longer a choice: She had to leave. For someone in such a prestigious position (head of college) taking an unplanned sabbatical just wasn't done. She worried about the negative repercussions on her career, but also knew something big had to change. She pushed hard and agreed to take time off without pay.

She and her husband discussed where to go. In theory, they were free to go anywhere. They didn't have kids they'd have to uproot. His job was flexible, and she could run the podcast from anywhere; they could be digital nomads. They looked for affordable spots all over the globe. They thought about Buenos Aires and Mexico City. And then Laurie recalled a film class she'd taken in college where the professor discussed a genre of movie he called remarriage comedies: a specific type of romcom where the couple is destined to be together, but breaks up at the climax. The trope in these movies was that one of the characters would depart the usual setting of the film—in a lot of classic films, New York—and decamp to a more rural location—Connecticut, for instance—to have the conceptual distance to realize that their true home was with their partner. The professor referred to these locations of epiphany as "green spaces" but acknowledged that green spaces didn't always have to be green; they just had to be significant to the person in some way, a place of respite.

When Laurie thought about a green space for her—a location that meant something to her and felt like a landing pad to rest on—she realized it was the town where she'd spent her young adulthood and met her husband: Cambridge, Massachusetts. Compared with South America, it was practically a stone's throw away from her life at Yale. But compared with how it *felt* to be there, versus how it felt to be in New Haven, there was a universe of difference. And for Laurie, it wasn't about getting *away* from Yale as much as it was about the effect this new place, her "green space," might have on her.

"That was really the goal in going, in getting the distance," she says now. "It wasn't just to get away. It wasn't to avoid. It was to get . . . the *right* effect."

Leaving New Haven certainly scaled back her overwhelming workload, but even with teaching off her plate, she was still working full time on her growing podcast, doing interviews, giving talks. According to Laurie, it was the change of environment that finally shifted her perspective and then shifted her emotions. The location allowed her to reconnect with parts of her old self and, in doing so, the longer trajectory of her life. She woke up in her tiny third-floor sublet, seeing the tops of the trees out the window, feeling as if she were safe up in a bird's nest. The smells and sounds of the city pulled her back in time to an era when she was happy and free and the future was full of possibility. She realized the future *still was* full of possibility. She wasn't locked into any of the choices she'd made that had landed her at Yale. She could choose that life, or she could choose something else.

New Haven, Laurie said, had been suffused with assumptions: about what was important, what was required, how things had to be. They had been impossible to shake off—like horse blinders. Within a week of moving to Cambridge Laurie had a revelation: Maybe she didn't want to be head of the college anymore. Maybe she didn't have to be; there were other options.

With her newfound clarity, Laurie could see that scaling back and stepping down from the role she loved was always the solution she needed. But she just couldn't see it from the space she was in. While the move and sabbatical were by no means easy or cheap solutions to an emotional problem, for her they were necessary ones that allowed her to regain her sense of balance and control, and ultimately move back to Yale following her sabbatical to continue educating students the world over. Sean, meanwhile, tapped into the same strategy in a less drastic way: a five-day visit to his resonant place was a powerful reset button.

When we change spaces, we *change the information that's coming in.* When you leave your house and walk into the nearby woods or park, your sensory experience shifts. Your attention shifts. And your perspective can shift as well. Of course, we don't always have the luxury of changing our environments when the going gets tough, even if we want to. Sometimes you can't retreat to your bedroom or head to your special place. You have to sit in your fluorescent-lit cubicle and get that report in on time. While there are plenty of constraints around changing your environment, we have a lot more control over our spaces than we think. The answer isn't always moving ourselves from one place to another. After all, humans have literally transformed the physical world over the past thousands of years. We have built bridges, highways, skyscrapers, golf courses, oyster farms, community gardens, and on and on. We have a knack for changing the spaces around us.

Why not change them to suit our emotional needs as well?

The Wisdom of Pizza Doggie Bags

A few weeks before I sat down to write this chapter, my wife and I hosted a football-watching party with a bunch of our friends. After stuffing ourselves, laughing, and yelling at the television for a few hours, everyone started gathering their things to roll home. It was at that moment I remembered something critical.

The pizza.

We had ordered way too much of it. There were two pies left, and the thought of eating them for lunch and dinner over the course of the next week both thrilled and horrified me (and, I imagined, my physician, who had recently cautioned me to eat more healthily) in equal measure. I knew something had to be done. So, I scurried into the kitchen and quickly boxed up a pizza doggie bag for everyone. As they were headed out the door, everyone demurred, not wanting to accept the going-away gift, but I insisted. No was not an option. And just like that, my friends were gone. And so was the artery-clogging (but delicious!) pizza.

Leftover pizza may not seem like an environmental or emotional hazard. But when it's staring at you from the refrigerator every time you open the door, it is. If it's in your space, it can influence how you feel. In my case, the emotional experience I was trying to avoid was *intense* desire upon seeing said cold pizza in the refrigerator later that night or the next morning or, honestly, any waking moment, followed by guilt and shame for consuming it in place of my steel-cut oats. Emotions that I knew I would feel if I succumbed to temptation and ate like my college student self for the next few days.

Here's the bottom line: If the environment is driving an unwanted emotional response, you can often change it by using a tool called situation modification.

There are countless ways we can directly manipulate our surroundings to shift our emotions. You can change them a little bit or a lot. In my case, I removed a tempting feature of my space (pizza) to shape my future emotional state. And the next day, guess what? I didn't have to perseverate between a slice and a salad. There was no late-night itch to sneak down to the kitchen for just *one more* piece. I *preemptively* foreclosed on all those unwanted feelings.

In one of the first-ever naturalistic studies of how high school students control themselves, the University of Pennsylvania psychologist Angela Duckworth and her colleagues found that when students are presented with hypothetical scenarios that require them to exert self-control, they overwhelmingly say that modifying their situations would be effective. For example, when asked how to prioritize studying over texting, one student wrote, "I would shut off my phone and put it under my pillow so I wouldn't be tempted to touch it."

Additional studies performed by Angela and her team showed that when students used these same situation modification strategies in their own lives, they made a real difference. They improved their ability to study effectively and accomplish their study goals over the course of a week compared to students who were told to use their "willpower" to resist their urges to not study or to students who were not given any instructions about how to manage their emotions. Other studies back this up: Successful students often change their spaces to accommodate their learning needs, cutting down on distracting temptations and boosting their focus.

Maybe these environmental tweaks seem like minor things— eliminating temptations like cold pizza or a glowing phone from your space. *But they are not minor.*

They can be huge.

Think about it this way: The long-term consequences of consuming certain types of food regularly can be quite serious. And if you have a personal goal to eat better (as I did) and you fail, because you set yourself up to fail, the emotional ripple effects can be significant.

And it's the cumulative effect of the spaces we are in most routinely that we want to become aware of. What is taxing you, draining you, tempting you, pulling you away from the things you want to do, the person you want to be? What in your environment is getting in the way of the goals you have for yourself? Maybe having a TV in your bedroom is working against your desire for better sleep. Or perhaps the lack of storage in your basement and the unrelenting clutter frustrates and distracts you every time you go down the stairs to work out. Instead of getting the full forty-five minutes in, you inevitably end up reorganizing the kids' overflowing art supplies and shortening your workout by a good twenty minutes.

Goal achievement is *tightly* intertwined with our emotional lives. When we have goals—like to be healthier or do well in school or work—and we don't meet them, that elicits frustration. We can get angry at ourselves. Ashamed. Filled with fear about the future. Sad at our inability to change the situation. When we *edit* our spaces to remove temptations, distractions, desires that are yanking on our attentional levers, we're short-circuiting that entire cascade of emotional dreck. Distractions in our immediate space may seem like a minor factor, but they are in fact big struggles with real consequences, and they are *eternal* struggles.

What is arguably one of the *most* famous emotion regulation stories of all time? Hint: It's the story that kicks off the most widely printed book that has ever been published.

The Bible.

It's a story about temptation, the story about Adam and Eve: *Don't eat the forbidden fruit.* And it's a story about the failure to resist that temptation. We still go back to that story today, thousands of years later, because it speaks to a core human truth: We are often drawn to things that aren't the best for us—physically and emotionally. We can keep fighting our temptations, and we certainly have tools to help us do that. Or we can just . . . remove the fruit.

I like to put the question to my students: If I remove something from my space so that I don't have to resist temptation, is that self-control? Consistently, half of them say no, it's not self-control, because if the alluring item does not exist, you don't have to exert any. But here's what I say to them: That's a narrow (and I'd argue incorrect) view of self-control.

Here's why: There's a pervasive myth that self-control is, by definition, an *internal* force (it's often referred to as willpower). You have to activate it to make it work, and it has to be hard. That's not right. Self-control can be easy. And one way to make it easy is to control your environment first, before you even have to deliberately activate forces inside you. Managing your emotions can begin *externally*. And if you're not doing this— removing triggers, temptations, and distractions from your spaces—you're not availing yourself of an effective tool for emotion regulation.

Which brings me to the next space modification: *adding*.

To shift with our immediate space, we're either *removing* something from the environment that is driving unwanted emotions or *adding* something to the environment that cultivates the emotions we want to experience.

Take, for example, one of the simplest interventions I've ever played a role in designing. We were interested in exploring novel interventions to help people manage emotional pain. So, we asked people to think about a really upsetting experience, like a breakup, a failure, a rejection, a betrayal, a humiliation. Then we showed half the participants an image of their mother, a person they were "attached" to, which we collected before the experiment, to see if looking at it would help them recover emotionally from dwelling on their negative experience. The other half of the participants saw the photo of their mom before we asked them to think about their negative experience to see if it would buffer them from becoming upset.

What we found: The buffering effect was nonexistent. But the restorative effect was substantial. Participants bounced back much quicker from those negative feelings. And it didn't only improve how people reported feeling. In another study with romantic partners, we found that when we asked people to just sit and write in a journal, stream-of-consciousness style, the ones who'd been shown the photo of their partner engaged in much less negative thinking than the control group as well. It was an exciting finding: Just *glancing* at a photo of someone you love improved people's ability to manage their emotions. And it's one that had a personal effect on how I structure my space. Shortly after completing this work, I went on a picture frame shopping spree. My office now is full of photos of my wife and kids, and other close family and friends. The first time my wife came into my office, she thought I'd gone a little crazy. I had to just tell her: The science backs me up.

Emotional Refueling

We spend a lot of time in our homes, and for some people, home is also a workplace. How we design those spaces affects our emotional well-being in sneaky yet important ways. So try this: Do a spaces audit right now. All it requires is for you to look around you the next time you are at home, in the office, or any other place where you spend significant amounts of your time. Think about how that space affects you and consider:

- What can you do—right now, today—to curate your space a bit to shape your emotional experience?

- What can you remove to turn *down the volume* on an emotional response such as stress or anxiety?

- What could you add to turn *up the volume* on a different one, such as calm and joy?

- What local, easy-to-access places would you identify as your own personal "emotional oases"? And how can you fold pit stops to these locations into your weekly or even daily routine?

Emotional oases are everywhere. They might be across the globe on a tropical island (wouldn't that be nice!), but they also might be across town, across the street, across the hall in your own house. We just tend not to plan our days around them. Think of yourself like a runner in a marathon. There are refilling stations all along the way that offer emotional restoration. How mindful are you of them? Are you accessing those recharge stations?

Meanwhile, the route you take through your environment turns out to have other important emotional side effects. When I was a junior in college, I lived off campus in a beat-up row house in West Philadelphia. Early one winter evening on the first day of the semester I closed the front door to my house and noticed a cute girl walking in front of me. I was running a few minutes behind on my way to class, so I fell into step about ten feet behind her. My mind was preoccupied with what was ahead of me that week, but eventually I started to notice that every turn the girl in front of me made was a turn I had to take too. After about ten identical turns, I started to get self-conscious, thinking she would worry I was some creep who was following her. But before long, we arrived at the same building on campus for a psychology course we were both enrolled in. After the lecture, I introduced myself. It turned out she lived in a row house two doors down from me. We walked home together that evening, and for the rest of the semester we met up going to and from class.

This may be the world's most pedestrian meet-cute (pun intended), but twenty-four years later I'm sitting here with that very same girl sipping her morning tea three feet away from me. I married her.

My connection with my wife, Lara, has enriched my emotional world in untold ways; healthy relationships are one of the best predictors of happiness. Yet the people who make up your closest relationships are often a function of your surroundings. Proximity is one of the most influential factors determining whom you marry and are friends with. Scientists call this phenomenon the propinquity effect—it's a powerful reminder that seemingly random features of our environment, such as who happens to live around us, can have a profound effect on the most important elements of our lives.

The places we occupy shape our emotional lives. They also play a huge role in determining the *people* who come into our lives and the intimate relationships we eventually build. As much as our environmental surroundings are constantly shifting us, so too are the humans who occupy those spaces: the people we are most intimate with, the ones we interact with in passing, and even the ones we meet only in the digital realm. As we'll see next, these interactions influence our emotional lives in profound ways, at times completely outside our awareness.

Chapter Seven

Catching a Feeling:
Relationship Shifters

"I want you to be an asshole."

As a professor who believes in cultivating a respectful class-room environment, it's not often that I start class with this instruction. But that's exactly what I said to a small group of executive education students during a workshop on managing emotions in teams. Before class, I had looked at the student roster and noticed that plenty of high achievers had enrolled. There were Special Operations military personnel and finance and auto industry executives, among other emerging leaders. They were being groomed for leadership roles, which meant they already had some of the qualities that businesses look for; they were ambitious, linear thinkers who were keenly aware of the need to collaborate. When I walked in and without even introducing myself asked for volunteers to lead breakout groups, I wasn't surprised that plenty of hands in the room shot up.

I led the lucky winners to a breakout room off the main classroom. Lining them up, I counted the volunteer leaders off, 1-2, 1-2, 1-2. Then I gave them the key instructions.

I told them that the 1s were going to be "cheerleaders."

The 2s were going to be "assholes."

There was a brief punch of nervous laughter before I continued to explain the exercise. I told them that each group sitting back in the classroom was going to be given a difficult logic puzzle that they had to solve in five minutes. When the team leaders went back to their groups, their jobs were simple: give Emmy-winning performances. The cheerleaders were supposed to act demonstrably supportive throughout the exercise. They were to lay the positive feedback on thick and maintain a high level of enthusiasm. The cheerleaders nodded. They were aspiring leaders; this was an easy lift.

Next, I told the assholes that they were supposed to be, well, assholes. Their role was to be stern and stoic and not give positive feedback. They were to glower and roll their eyes when someone asked a question or suggested a solution. Before I even finished explaining their role, I saw looks of dawning horror on their faces. After all, they had just been getting to know the other students in their cohort. Their desire to belong and to be liked was triggering an emotional response. What I asked them to do violated social norms in a moment when they were trying to build relationships. As I continued to lay out the rules, members of the asshole group looked increasingly uncomfortable, squirming and sighing and shifting from one foot to the other. With assurances that the deceit would immediately be revealed to the class at the end of the exercise, they complied and walked into the classroom.

Within seconds, stark differences emerged between the two

groups. The groups led by the cheerleaders were loud. They were talking, laughing, making suggestions, and moving around the table. They were spontaneously complimenting one another for creative ideas and literally patting one another on the back. As for the asshole-led group, the atmosphere at their tables suggested they were breaking rocks in a gulag. Their heads were downcast and their faces were grim. Everyone was quiet. Occasionally, someone would utter a brief, almost contrite suggestion, but very quickly the silence would resume. No one was making eye contact, and they weren't getting anywhere; none of these groups solved the puzzle.

Later, several students would tell me that this exercise was their favorite of the four-day workshop. Despite the temporary social unease, this group of future leaders felt that it communicated viscerally something they only knew intellectually: just how quickly emotions can spread.

Like a Virus

The rapid transmission of emotion can change how those around us feel, think, and behave. It's how one student with a bad attitude can derail a carefully constructed lesson plan. Or how one team member who has a problem with the project can grind progress to a halt. And the virality of positive emotions is just as potent. It's the person at the staff retreat who takes it upon themselves to turn up the music, start a dance party, and change the course of the day. Or the emergency responder who arrives with the calm and confidence to instantly alchemize a situation, taking it from crisis to manageable challenge in mere minutes.

Like a rapidly circulating virus, *emotional contagion* is what happens when we "catch a feeling," and is something that our species has been taking advantage of (and suffering from) since time immemorial. The intricacies of how emotional contagion works on a psychological and neurobiological level are complicated, but it starts very simply: with *mimicry*. If you've ever watched a baby break out into a smile when you grin at them, you've seen this firsthand. Whether it's your tone, the look on your face, or the way you're tilting your head, other humans in your presence pick up on these expressions unconsciously and, to varying degrees, respond through mimicry within seconds. When we mirror others in this way, we often start to feel the same way they do. In the moment, you are likely completely unaware of it, but the physical movements you mimic transmit signals to your nervous system that set off an emotional cascade within you.

Reams of research indicate that catching feelings like this can influence how we feel about the work we're doing—our commitment and satisfaction, our ability to empathize, our susceptibility to burnout. Emotional contagion shapes how we handle conflict, how we coordinate with other members of the group, and how well negotiations go. It plays a central role in our day-to-day lives, including much of the time we spend in the digital world, which provides us with even more opportunities to shift other people and be shifted *by* them.

Consider the way emotions can spread through social media at a pace and reach unprecedented for our species. A video is posted. It resonates. It's shared once, twice, 4 times, 16 times, 256 times; it keeps going. Twelve days later, that video, along with other related ones, has been viewed 1.4 *billion* times and has spawned a social justice movement that persists to this day. I'm talking about the cell-phone video of George Floyd's murder, which spread outrage around the globe. Think of any social media movement that has been influential—the Arab Spring, Black Lives Matter, the Me Too Movement—and you can see that emotional contagion was a wave upon which it rode.

It's easy to overlook the effects of emotional contagion because it happens so quickly, and largely through psychological forces outside our awareness. But it's also because many of us think about our emotions as private experiences and overlook the fact that our minds are permeable. Emotions from the inside flow out and emotions from the outside flow in. Far from being siloed within us, our emotions are deeply social; other people can and do tug on our emotions, activating and deactivating them, turning their volume up or down, or transforming our experience of them altogether. And this is true whether we're talking about someone in our inner circle, such as a partner, child, or close friend, or the other people that simply cycle through the background of our days, out there in our broader communities and in the digital scroll of our social media feeds.

We are a deeply social species. We routinely turn to other people to help shift our emotions—sometimes intimately, in conversation with a confidant, and sometimes at a distance, taking in what we can of other people's lives and struggles, to evaluate our own. And *how* we do these things shapes whether our interactions with others during times of emotional need help or hurt. The conversations we have, and how we have them, can lead us to ruminate and harmfully avoid, *or* they can allow us to reframe, positively distract, and soothe our nervous systems. This may often begin with emotional contagion—other people's emotional states bleeding into ours—but it doesn't stop there: what we talk about, how we talk about it, and how we *think* about and interact with the people we know have the potential to shift us. But we don't have to leave those factors up to chance. There's a sea of humans around us—some intimates, some mere digital acquaintances—and we have a say in exactly *how* they affect us.

There are, of course, myriad ways that interacting with others can shift our emotions. But there are three in particular that offer a "high return on investment." In other words, they pack a big emotional punch. So, in the pages to come, we'll focus on these specific levers within our relationships that can easily take us off track, and how we can instead use them to shift in the right direction.

"Talk to Me"

Helping each other through dark times is one of the most important things we can do. But we don't always get it right. Sometimes, when you go to someone else for help with an emotional issue, you find that what unfolds ends up doing more harm than good.

In my first book, *Chatter,* I spent a chapter looking at the role that other people play when we get stuck ruminating in our heads, locked in a cycle of negative self-talk. We looked at how in moments of distress research shows that cross-culturally people seek support from other people. Despite the obviousness of our need for each other when times are tough, this continues to be an area where people struggle. As many of us have no doubt experienced, sometimes other people really help. But other times, we leave those conversations feeling just as stuck as we were before, and too often even more inflamed because of them. As a result, our conversations with others can shift us either way—toward helpful insight or unproductive co-rumination.

There are two key ways other people can help us when we go to them for emotional support: They can satisfy our core need for empathy and validation, on the one hand, and they can help shift our perspective, on the other. We routinely balance this formula incorrectly, failing to strike a balance between these two critical elements when we try to either get help from others or provide it.

Most of the conversations people have with others about their emotions take the form of them rehashing their feelings. This reflects the fact that many people think that getting good support begins and ends with spilling our guts to someone who will listen. That's not what the research shows. Too much time spent on rehashing a problem is called venting or co-ruminating, and it can make things worse. It's the well-meaning friend who is excessively validating of your point of view, digging you deeper into the very outlook that is making you unhappy. *He's SUCH a jerk. You should be outraged. If I were you, I would tell him off.* And it's also the friend who asks you probing questions, digging for every detail, and in doing so causes you to re-live the emotions that have you worked up in the first place.

All that venting and validating can be good for strengthening your relationships; it's nice to know there's someone who is willing to take the time to listen to you, and it feels good to engage in those divulsions in the moment. If you're wondering why that is, research suggests that when we are threatened, early attachment instincts activate that propel us to go to our support providers, the people around us who care about us, to receive social and emotional support. The idea that we're stronger with other people is deeply embedded into our psyches. It's a primitive instinct. We're recruiting them, in a sense, to be on our team.

And it feels good to share things about us with others. In 2012, the Harvard social neuroscientists Diana Tamir and Jason Mitchell published a landmark paper showing that people value disclosing their thoughts and feelings. Doing so consistently activated the brain's reward circuitry. That's right: The same "feel good" dopaminergic pathway that we associate with pleasurable experiences such as ice cream and sex was activated when people were given the opportunity to talk about themselves. People were even willing to forgo money in lieu of disclosure: It was *more rewarding* to disclose something about yourself than literally be paid.

So, the motivation to express runs quite deep. But it's not enough. If that's all you do in a conversation, it can cause the wounds you're talking about to fester instead of heal. A better strategy is to get it out, but *then* talk about the problem in a way that helps you put it in perspective, making use of the other person's fresh point of view. If you've ever been stuck in an emotion, you know that it can very quickly result in tunnel vision. All you can focus on is the source of the problem or the pain that you're in. When this happens, other people can be crucial for helping us shake off the blinders. Perspective shifting is a lever for emotional change that is already within us—and one we just talked about two chapters ago—but taking in a different perspective during an emotionally charged moment can be hard. When someone we trust nudges that lever a little from the outside, it can be exactly what we need to grab hold of it ourselves.

There is of course an art to doing this as the listener. You have to consider the order of operations. People need to "get it out" first before you can help them think about the problem differently. And the more intense the emotion, the more time it will likely take before the person is receptive to having their perspective broadened. When people are upset, they want to feel heard, understood, and validated. That's essential. But if you *only* do that, as a friend, colleague, or loved one, you are missing an important piece of the puzzle. A healthy perspective shift happens only if you add in the *cognitive* support as well, meaning that you help the other person work through the situation instead of just commiserating. You offer up potential solutions and perspective shifts that might help someone see their problem in a different light.

The key is to both empathize *and* help them think through their issue from a broader perspective.

The same goes for what you are asking for from the people you talk to when you are the one who needs help. If you sense the conversation is veering too far into co-rumination, you can take control and initiate a turn toward more concrete help by asking them what they would do in your situation, or what it looks like from the outside. These are, of course, some of the ways that trained therapists practice certain types of empirically supported interventions to help their clients. But not all the issues we struggle with require therapist intervention, nor is it always practical or even accessible for everyone. The point isn't to replace professional counseling with a chat with a friend; it's that we can all use these basic principles of emotional exchange a lot more effectively, *with or without* therapy in our lives, if that's what we need. Facilitating an emotionally deft conversation can feel like trying to dance your way nimbly through a minefield. But there are two simple guideposts you can use to steer the exchange.

When *you* are the one in the "emotional adviser" role, start off with listening, empathizing, validating, normalizing. Try: *Tell me more about that. That sounds hard. What do you think about that? I'm not surprised you feel this way.* But keep an eye out for any signs that you might be veering into co-rumination, such as you're getting just as wound up as they are, absorbing the anger, sadness, and so on, and amplifying it.

When you see your opening, start to gently zoom out and see how they respond. If they're not quite ready, you can come back to listening. And if you're unsure, just ask! Say, *I have a thought on this, are you okay to hear it?* They might say yes. Or they might say no—they just want a listener. And that's okay. Not every emotional conversation has to end with a solution. Sometimes it's enough to just be there and hear them in that moment. You can usually come back and revisit the conversation, providing you with additional opportunities for zooming out.

And when *you* are the one who needs the emotional support, *know whom to go to.* When you really need help, you need someone who can do the two things above: first listen, learn, and empathize; and then transition to helping you shift perspective. *Not everybody can do this.* I'm confident that there are many wonderful people in your life right now—people you love deeply, whom you value and go to for any number of your relationship needs—who are simply not meant to be your core emotional advisers.

Here's how I encourage people to identify their emotional advisers in the classes and workshops I teach, and I'll invite you to try this right now, if you like.

Take out a piece of paper and draw two columns, labeled "personal" and "work." List all the people you go to to talk about problems in those two domains. You may have some of the same names in both columns, or you may have no overlap. You may have a long list, or very few. That's okay. What we want is a rough snapshot of your support network, whether it's just a couple of people or a crowd.

Next: Think about the balance of venting and zooming out. Some of the people on your list might be fantastic candidates for allowing you to just vent, and venting can be *great* for relationships and bonding. It's just not as great for working through the actual problem. Meanwhile, people who veer hard toward "advice giving" from the outset can be tough to talk to; that can be bad for relationships. In a recent workshop, I informally polled the group and asked them, "What do you call people who jump right into giving advice without first taking time to listen and learn about what you're going through?" The common response was, "Jerks!" One person yelled from the back of the room: "Husbands!" (As a husband, that one stung!)

Finally, go through your list and circle the names of the people who do *both* things for you: *They listen, and they broaden your perspective.* These are your emotional advisers. As for the rest of your network? Not everybody can be everything to us, and that's okay. You can let people into your world without necessarily turning to them for help shifting.

Another option? Educate them. If they're a co-ruminator, clue them into the research by saying something like, "I came across this interesting chapter about how venting doesn't actually help in the long term. I had no idea!" Or if they're an advice giver, front-load it: "Can I just vent about this for a bit, and then maybe you have some ideas for me?"

In the same way a start-up carefully vets its trusted advisers, we should think carefully about whom we are recruiting to help coregulate our emotions. Going forward, keep in mind: When you go to someone for support, are they good at letting you express your emotions, but not for too long? Are they unafraid to give you the straight talk when you need it? Do they usually bring up perspectives that you haven't thought of on your own? Answering these questions is just one way of seeing exactly who checks the most wise-catalyst boxes. If there are friends who fall short, this doesn't mean you have to get rid of them. It just means that you know more about their strengths and weaknesses. Successful businesses spend a lot of time thinking about who sits on their own advisory boards—so should you.

Finally, remember that these kinds of "emotional advising" conversations are not the only way to help someone in your life shift, or to lean on them to help *you* shift. Positively shifting someone else can be done either with or without a person's awareness; we can also shift those around us by grabbing their attention or activating their senses. This can be useful in delicate situations where someone might push back or get defensive if we try to talk to them about their problems without them asking for support. Whether it's encouraging someone to go to a movie for an emotional reprieve or triggering a shift in someone we care about by engaging their senses, we all have the capacity to activate other people's attention and sensory shifters. When my daughter was having a hard time shifting out of a bad mood before her soccer game, I turned on Journey. The music I played was able to activate her own internal sensory shifters. I hit Play, but she listened.

Now, there's another way that other people shift us emotionally, and they're often doing so completely unintentionally, without any awareness, just by existing. And how we engage with their existence provides us with a golden opportunity to sink into despair or lift out of it.

Harnessing the Thief of Joy: Comparing Wisely

Not too long ago, on a cold winter night, somewhere in America, there was a tween who wanted more screen time and a parent who said no. The tween in question was advocating vociferously for her right to scroll for an extra thirty minutes at night. The parent in question argued against this request by trotting out the fact that none of her friends' parents let *them* have screens after nine o'clock.

"I thought we don't compare ourselves with other people in this family?" the tween said in reply.

The parent in question—who was me, by the way—just got served (that phrase might date me, but oh well). Because what my daughter said was true. Since they were young, I have told my daughters not to compare themselves with other people. I have argued countless times that comparisons are the "thief of joy." They put you in a box, limiting your aspirations and creativity to what you see in others.

Although my daughter didn't prevail in her quest for more screen time, she did help expose one of the shoddiest pieces of advice I have ever given. In my defense, I did what we've all done before, which is repeat received wisdom without delving into the nuances. But now is the time to set the record straight, which starts with interrogating the idea that all social comparison is unhealthy. (Spoiler alert: it's not.)

There are, of course, good reasons people are wary of comparison making. As so many encounters with Instagram influencers validate, social comparisons can be harmful. Some of my initial research on social media sites in the early 2010s showed that interacting with Facebook (the dominant platform at that time) negatively affected well-being. It was a straightforward finding: The more time people used Facebook, the more their positive mood declined over time. It wasn't until a set of follow-up studies that we understood why this was happening: The more people scrolled on Facebook, the more envious they felt, which in turn predicted declines in their well-being.

Roughly half a decade later, in one of the largest studies on social media and social comparison ever performed, a team of scientists from Facebook corroborated these results and extended them. In a study of 37,729 participants, more than 22 percent of people said that they had compared themselves with others on Facebook and felt badly about it during the past two weeks. More than a third of that 22 percent said that those negative feelings lasted for a day or more. As my colleagues and I have noted, if you extrapolate from these results, considering how many people use social media platforms such as Facebook (three billion people, in that case), you can approximate that hundreds of millions of people feel bad every day because of digital social comparison.

This explains a lot about the current discourse around social comparison. Because most of us spend a lot of time on social media (about two and a half hours a day), social media sites become an obvious purveyor of unhelpful comparisons. And because negative experiences loom larger in our minds than positive ones, it's easy to understand how we can fall into the trap of thinking that if this type of social comparison is harmful, then *all* social comparison must be bad. Sometimes even the very people who should know better fall into this trap (ahem).

But before we write off social comparisons as inherently toxic, let's remember that it's a universal feature of human psychology; we can't help but compare ourselves with other people. It's a behavior that is baked into our brains. Once we start making comparisons around the time of preschool, we keep doing it at every age. And that's true regardless of our income bracket or culture. It's inescapable; we are constantly weighing how we are doing against others. Sometimes we do this spontaneously, like when we're mindlessly scrolling through Instagram. And sometimes we do it deliberately, like when we're scanning the alumni magazine announcements to see how we stack up against our college nemeses.

We engage in these comparisons so often for a very simple reason: *They help us make sense of ourselves.* Our self-worth isn't just determined by how objectively good we are at sports or how many A's we get in school; it's also about how we fare in comparison to others. When there are no clear objective standards (Am I smart enough? Good-looking enough? Social enough?), we look to others to form our opinions and direct our behavior.

To be clear, social comparison is not just a rudimentary status ladder, with one metric determining who is "above" or "below" you. When we compare ourselves with someone else, we can hold the duality that we may be more skilled than they are in certain arenas, while less skilled in others. That they may have certain resources we lack, and vice versa. That they might have achieved more career success than we ever will, but lag behind us in terms of relationships or family life. Social comparisons are a nuanced psychological apparatus that help us crystallize and differentiate our goals, values, and sense of self.

And we're constantly using this tool, often without a second thought. But as our experiences on Instagram and countless other circumstances suggest, it often gets us in trouble. One of the largest analyses of social comparisons to date reviewed more than sixty years of research on the topic and found that most of the comparisons people make are to people who are outperforming them in some way, and they generally result in them feeling bad.

But here's the point we often miss about social comparisons: They can be harnessed for our betterment if you understand how they work. For a study Micaela Rodriguez, Ozlem Ayduk, and I performed, we recruited participants who were dealing with a difficult problem and then asked them to think about someone they knew who was faring worse. We found that many of the participants found comfort and strength in those comparisons; they felt more optimistic and less negative than another group of participants who were just asked to reflect on their problem. One student reported thinking about their family members who were grieving and suffering, but who still managed to get out of bed every day. This comparison sparked the realization in the student that they were actually pretty lucky and gave them a boost of motivation: *If they can go out every day and do what needs to be done, with everything they're dealing with, then I can too.*

• The social comparisons we make—ones that lead us to feel good *or* bad about ourselves—are vital to our ability to thrive in much the same way that both positive and negative emotions contribute to our well-being and success. So first, let's do away with the notion that all social comparisons are bad. This in and of itself can pile negative emotion upon negative emotion. If you find yourself feeling insecure and envious of a friend who seems to have achieved things in life that you're still struggling toward, no need to get down on yourself further by castigating yourself for making comparisons. We have to let that unrealistic expectation go and understand that like intrusive thoughts, or chatter, comparisons are likely going to happen. The real pro move here from an emotion regulation standpoint is to take something that you already do automatically and leverage it to your advantage.

So here's the simple scientific crib sheet to leverage the people around you to shift in the direction you want. Once you know it, it becomes easy to amplify the comparisons that serve you and ditch the ones that don't.

Comparing yourself with someone who is outperforming you could result in feelings of envy and dejection if you focus on the things they have and you don't, or it can be energizing and inspiring if you use these comparisons as a source of motivation, for example, "If they can achieve that, so can I."

Comparing yourself with someone who is doing worse than you could result in fear and worry if you think about how you could fall into similar circumstances—"Uh-oh. If it happens to them, it could happen to me"—or it can elicit feelings of gratitude and appreciation if you use that comparison to broaden your perspective, for example, "Wow, things could be much worse; I'm doing great."

Using this tool is like using mental time travel: When uncorralled, it can drag you places you don't want to be. But when it is used skillfully, you can hijack something your brain is *already doing by default* and turn it into an asset.

What I wished I taught my daughters earlier was this very nuance. How we feel about ourselves hinges not just on *whom* we compare ourselves with but *how we think* about that comparison. Now I remind my daughters that comparisons are one way we learn about who we are and how we stack up. And that they can be misleading or harmful depending on how we engage in them. And I talk to them about frequency. If you engage incessantly in comparisons to people you assume are doing better than you, they can start to undermine your well-being. That's when it's smart to pull your attention away from comparisons, focusing on them only when you need extra help shifting your perspective. And finally, I remind them that the information we get from comparisons is just *one* data point. These comparisons are context-specific judgments. We are many things to many people; others may compare themselves with us in ways that would shock or surprise us. There are different contexts, different people in your life that you might call to mind, depending on how you want to shift your emotions.

If you still have any hesitation about using social comparison as a tool, just remember that *this is how we work*. We are a hierarchical species. Social hierarchies exist, and we exist in them. And if you're still feeling bad about feeling better by comparing yourself with others, there's something you can do about it.

Taking Care

In what must have been one of the most fun psychology experiments of all time, the researcher Liz Dunn and her colleagues walked around the campus of the University of British Columbia in 2008 giving people envelopes of cash. But before the researchers let them look inside, they had the participants rate their current levels of happiness. With that formality out of the way, the lucky individuals were randomly assigned to one of two groups—the "personal spending" group or the "prosocial" group.

The people in the personal spending group were told they had to spend the money on themselves by that evening. They could do whatever they wanted with it as long as it benefited *them*. For example, buying an ice cream, paying off an overdue library fee, or going to the movies. Those in the prosocial group had to spend the money on *someone else*, also in whatever way they wanted, whether by donating to charity or taking a loved one out for coffee. Each envelope held either a five- or a twenty-dollar bill. After the 5:00 p.m. deadline that same day, Liz and her colleagues followed up with participants in both groups and had them rate their happiness again.

Many of us think quite a lot about money and happiness. There are countless headline-grabbing studies on the topic, but most circle around the question of how *much* money we need to be happy. What made Liz's study so interesting is that they didn't focus on quantity of money as the linchpin of happiness. Instead, they questioned an intuitive belief: that spending money on ourselves feels better than giving it to others.

What they found was exactly the opposite.

After crunching the numbers, Liz's team discovered that the people who spent money on others, regardless of the amount, experienced a significant boost in how happy they felt compared to the people who spent money on themselves.

Like forty-niners bum-rushing the Sierra Nevada in search of gold, other scientists sought to build on Liz and her team's discovery and confirm that *whom* we spend our money on is an important predictor of our emotional well-being.

One question about this effect was that it might apply only to small samples of well-heeled college students in rich countries. But it turns out the giving impulse taps into something more primal. Later studies confirmed that these principles operate in larger groups of participants living in rich *and* poor populations.

You also might think that when the personal stakes rise, people might be more inclined to help themselves. No again. Studies performed during COVID, when concerns about personal well-being heightened, have revealed similar results. In each case, happiness was improved when participants spent money on others instead of themselves. Collectively, these findings demonstrate what a powerful force for happiness *giving* really is.

In many ways, this is a surprising effect. Scientists and philosophers have long described people as selfish, increasingly concerned with privileging their own desires over the needs of others. But when you really think about it, as long as we have been living in groups, well before the advent of money, we have been caring for one another. Even when it is not socially required or even recommended. Sometimes we choose to relinquish our own resources and risk ourselves to help others. News stories abound telling of people rescued by their neighbors during natural disasters such as floods and wildfires, Good Samaritans who showed up to get people out or bring them supplies. Why do so many people go beyond simple gestures, like giving five dollars to a stranger, and put themselves at risk to help other people?

Because sometimes giving *is* actually in our own self-interest. When we perform acts of care, we are strengthening social bonds and increasing the likelihood that we too will be helped in the future. And because kindness is a perceived social good, when we perform it, we also gain status (giving makes you look good), which helps us find and keep friends and partners in our social circles. For all these reasons, when we help others, we feel happy. Evolution gifted us a squirt of joy each time we pitch in because, ultimately, it helps us perpetuate the species.

Study after study backs up the phenomenon of feeling better by doing good for others. In one meta-analysis on the effects of performing kind acts that included twenty-seven experiments and more than four thousand people, researchers found that doing nice things for other people significantly enhanced how good they felt and were as effective as other interventions such as practicing mindfulness and gratitude. These effects are true for people regardless of age, sex, or how prone they were to experiencing social anxiety.

You don't need to volunteer at the soup kitchen or start a philanthropy to tap into the emotional benefits of helping others (although those are great things to do!). It's as simple as recognizing when you could use an emotional lift and looking around to see who is nearby. Let's say your partner has a short fuse and seems prickly. You might get frustrated that they're taking it out on you, or you might notice they have a packed schedule and decide to take lunch-making duties off their plate. There's a mood-boosting benefit for *both* of you. Or perhaps your elderly neighbor has seemed down and reclusive lately. You look outside and see that their lawn is overgrown and untended, so you go over and mow it the next time you do yours and derive a sense of satisfaction knowing you made their day just a little better. In making these small gestures, we are taking part in a millennia-old tradition of mutualism, the idea that we benefit from relying on each other, which offers a sense of belonging and purpose that shifts emotions in its own right.

People Power

Our relationships are the deep currents of our emotional lives. They move us regardless of whether we understand them, but now we do. We know that feelings can be contagious like the flu. That other people can powerfully move the needle on our distress. That not all social comparisons are harmful. And that helping others helps ourselves. This knowledge is critical, because when you know, you can play with these levers.

If I'm offered a job and catch some negative feelings from staff during an office tour, that is valuable information about the kind of emotional contagion I might be exposing myself to every day if I worked there. In a conversation with someone I've turned to for advice, I have the tools I need to assess: Is this conversation helping me? And if not, I know how to get it on track, and when to bail out and find a different emotional adviser. If I know that it's possible to leverage social comparison positively, I have a shot at reframing whatever negative thoughts come up when I inevitably compare myself with someone doing "better" than me. If I know that helping others can help lift my mood, I can have a list ready to go of ways I can help my community. The beauty of this emotion-relationship knowledge is that it doesn't help only you. It will also help you be a good emotion-regulation partner to the people around you—your kids, your spouse, your roommates, your coworkers, your siblings.

If we are individual threads, and our relationships are like a braid, then the next logical step is to consider the larger weave. We don't live in a world of isolated, braided relationships; we live in a complex weave of relationships called cultures. Think about these cultures as the larger tapestries where those individual braided connections come together to form unique patterns of emotion. These larger tapestries become the constant backdrop of our lives. Everything that unfolds, including our emotions and how we respond to them, happens within this broader context.

Something as constant, as all-consuming, as *culture* can be difficult to see; we live so fully immersed in it that it can become invisible to us. And yet as we'll see next, it may ultimately be the most influential of any of the emotional forces we've discussed in this book.

Chapter Eight

The Master Switch:
Culture Shifters

One Friday, about five years into her sobriety, Hollis walked into work and ran smack into a crisis. She arrived at 9:00 a.m. with a mocha in one hand and a laptop bag in the other, but by 9:30 a.m. the coffee was gone and so was her job. Not long after, she was climbing the steps to her house, carrying a box with all her stuff, gobsmacked by an unforeseen layoff.

Until then Hollis had no idea the tech company she worked for was "restructuring," and since she didn't see anyone else with a box and a grim look on their face, she wondered if the company was just trying to soften the blow. Mercy layoff? Maybe Hollis hadn't been let go so much as fired? Either way, her inner critic had already fired up the self-loathing machine and was churning out zingers. *Of course, I got fired! It's only surprising that it took them so long to figure out that I couldn't handle it. Who did I think I was, pretending as if I deserved a job like that—with benefits, with people who had their shit together? Ha!*

When Hollis walked in the kitchen and saw yesterday's mail, replete with mortgage and utility bills, her heart rate picked up and her palms began to sweat. She had just bought the house last month, something she never imagined was possible only a few years ago, when she was in and out of the hospital, living with her parents, and frequently drinking herself into oblivion. As panic locked up her ability to move, she sat down on the floor of the kitchen and imagined what it would be like to be homeless. The thought of telling friends and family made her want to crawl into the bathtub with a joint and a bottle of bourbon. The fear made it hard to breathe, while the shame she was experiencing prickled her skin. Hollis couldn't remember ever feeling this emotionally disturbed as a sober person. And it was terrifying.

Respond, don't react.

The thought drifted into her head—something her sponsor, Ray, always said.

For an hour or so, Hollis sobbed and hyperventilated on the kitchen floor, but the words kept coming back to her, and so she tried to break it down and remind herself of what she already knew to be true. She had learned that a response was intentional and constructive, while a reaction was knee-jerk and emotionally driven, something that usually led to the bottom of the bottle. Hollis knew what to do instead. Last year, when one of her friends committed suicide, she had fallen apart. But then she called Ray. Then Genelle, then Arthur, then Susan, then Carla, then Franny, then Tyler, until finally John picked up. The first ten numbers in her contact list were all people from Alcoholics Anonymous (AA), their names followed by a little *a* so they'd show up in the same group.

Sitting on the floor in yet another emotional predicament, Hollis scrolled through the list and started dialing. She reached four friends that day and sat on the floor talking until well into the afternoon. Each of them reminded Hollis of what she needed to do to get through the day sober. Part of the reason she called AA friends first was the shared language and experience. Everyone knew what to say, knew what Hollis was going through, and they could help in a way that no one else could. On the phone with Ray, she planned out which meeting she would go to that evening. At the end of a call with Genelle, they recited the Serenity Prayer together, and Hollis promised to take ten minutes after the call to meditate. When Tyler picked up, he reminded her of the saying "I came for my drinking, but stayed for my thinking," which helped refocus her attention on disrupting the unhelpful emotions she was experiencing.

As Hollis had heard so many times before, people don't go to AA because drinking is their problem; they go because drinking is their solution. If Hollis wanted to solve the problem differently, she would have to think about it differently. Carla picked up the phone and reminded her of two key steps in AA: letting God take the wheel and relinquishing control. Hollis's tendency to project her fears into the future was something she was always called out on. So, when Hollis told Carla about her fear of winding up in a tent living under an underpass, Carla said, "Where are your feet?" Hollis rolled her eyes and groaned but took the point. They weren't there *yet;* don't borrow trouble.

The fact that Hollis was even able to pick up the phone that day and call people felt like a minor miracle. People in AA call it the one-thousand-pound phone for a reason; for those in distress, bubbling over with shame, it can feel impossible to ask for help. But Hollis had worked the steps and taken the strong suggestions of some elder members to get three numbers at every meeting and practice calling people. The thinking was that in times of stress and need you're more likely to follow through if you've practiced when you're not so out of sorts.

By the end of the day Hollis was let go, she had gone to a meeting, eaten nasty takeout, and curled up with her cat and wept, but she never took a drink. She got through that day, and the next day, and the next. She found another job. And despite the whirlwind of emotions that threw her around, she managed to keep her house and her sobriety. Looking back on it now, Hollis credits the rock-solid *culture* of AA: "It saved my life."

————

Culture has often been called the air we breathe. And in many ways, it's the most profound and foundational force in our emotional lives—which is why we arrive at culture last.

When I think about how culture fits with the other emotional shifters in this book, I picture a beautiful set of Russian nesting dolls my mom brought back for me after a trip to Europe years ago. I remember the wonder I felt opening the first doll to find another meticulously painted, slightly smaller doll resting inside. The emotional shifters we've been talking about throughout this book are a lot like those dolls: They live inside us in similar nesting layers, building on each other to affect our emotional lives.

We started this book looking at the innermost layer—the way sensation, attention, and perspective shift our emotions. Then we moved on to the spaces we inhabit, and how they affect those internal shifters; next, we expanded to the humans around us, and how those relationships move our emotions around.

Culture's impact filters down through all those inner layers, pulling our emotional shifters in ways we may not even realize. It influences the music we listen to, the food we eat; it shapes our philosophy about emotions—like whether they should be felt or suppressed—and how we make sense of our lives. Culture has a profound impact on the spaces we exist in, from the layout of our cities to how we build and decorate our dwellings, and it sets the rubric for our interactions with the other humans around us, how intimate we become with them, under what circumstances we spend time together, and what we expect of our relationships. This isn't merely a nod to the interconnectedness of life; if we can learn how to navigate the cultural containers we inhabit with an eye toward emotion regulation, we can change our emotional trajectory not just for ourselves but for others as well.

But like so many of the shifters we've talked about throughout this book, there's a catch: Culture's impact on our emotional experience can split hard in two opposite directions. Because it is the air we breathe, it can be either the foundation for emotional wellness or the reason for ongoing distress. When culture is strong, what we most need to do is understand which tenets of culture to grab onto to help us through. And when the cultural container is at the root of our struggles, those very same tenets become the levers we use to change the culture. So in this chapter, we'll look at both of these angles: how exactly to lean on culture when it's healthy, and what to do when it's not.

But first: What exactly *is* culture?

The Air We Breathe

When we think about culture, we often think about the ever-changing winds of art, music, fashion, lifestyle, and wellness. It's Taylor Swift's Eras tour, CryptoBros, the Kardashians, the carnivore diet, dad bod, the latest TikTok dance, tiny homes, Inbox Zero, and, at the time of my writing this paragraph, the *Barbie* movie (the fact that each of these cultural touchstones will be yesterday's news by the time this book goes to print is kind of the point). We also see culture as part of what defines a group of people, like Native Americans or the Amish, and most of us take care not to appropriate their traditional garb or ceremonies out of respect. We know culture plays a role in the myriad ways we eat, talk, celebrate, and mourn. It is half legacy, half living thing.

The very first culture we are welcomed into is our family of origin, with its own religious, ethnic, and national particularities. The next is usually our school's culture, and our chosen group of friends. Then, depending on our choices, it's sports teams or after-school clubs, and finally the colleges we go to and the workplaces we join as adults. Of course, culture exists at so many more levels than this—from Rotary Clubs to entire geographic regions. For every group of humans that gathers consistently, a culture forms. And when you closely compare the emotional landscapes of different groups, whether they be tribes, families, organizations, or neighborhoods, striking differences emerge.

Take corporate culture as just one example. A start-up that prides itself on a move-fast-and-break-things ethos might develop a culture that treats anxiety and stress as emotional experiences to embrace. In contrast, a decades-old company with a

sturdier financial foundation might prioritize employee reten-
tion, communicating to its staff the importance of work-life
balance and keeping stress levels in check. Depending on which
work culture you inhabit, how you feel about stress and anxiety
may vary—widely. And how we feel about emotions affects the
ways we seek to regulate them, or if we even try to regulate them
at all.

A lot of us try to get our emotions under control by *suppressing* or *avoiding*. As we've seen, in the United States these
tactics are often viewed negatively, while emotional expression
is highly valued. This isn't the case everywhere. In so-called col-
lectivist cultures (like Japan) where social harmony is highly
valued and people privilege the group over the self, individuals
are more likely to suppress their emotions and to view doing
so positively. In individualistic cultures (for example, European
Americans), social harmony is lower on the priority list. So,
people are more apt to express themselves.

No matter which group we belong to, culture helps us both
navigate the world and make sense of our place within it. It provides us with different expectations, vocabularies, and modes
of interacting. Its constituent parts have evolved to fit the particular demands of the groups we belong to over time, which
is why there are as many gradations of cultures in the world as
there are contexts, generations, and environments. A diversity
of cultures exists not only in the world but also within our own
lives. When I wake up and have breakfast with my family, I am
swimming in one culture. When I go to a university meeting, I
am swimming in another. Depending on the groups we belong
to, we could be dipping in and out of half a dozen cultures in
the span of one day.

Since culture can be both large (the culture of a nation) and small (the culture of a friend group), we can define it very practically by saying culture is the embodiment of ideas and practices that have been passed down over time through different pathways (for example, families and institutions) that give us the best chance at thriving within our particular context. You can think of culture as a kind of master switch in a "smart" home or hotel room. If you've ever used one, you know that with one press of a button the smart switch can control the temperature, lights, electronics, and blinds. In a way, that's what culture does as well. Passed down and evolving from generation to generation, culture shapes different facets of our experience within a group. What we think, how we act, and, crucially, how we manage our emotions.

Every culture has the ability to pluck our emotional strings, but some cultures are more powerful than others in terms of their ability to help us regulate our emotions. Hollis is an example of one of the millions of people who have used the culture of Alcoholics Anonymous as a tool not only for their recovery but also for reaching the emotional stability that is at the heart of it. This organization is a free, open "fellowship" for people who are looking for support for their substance abuse problem. Founded in 1935, this is also a group around which a powerful culture has formed, a culture that people often say has transformed their lives. In many ways it is like a boot camp for managing your emotions. Many people who walk into AA meetings have seen unchecked emotion ruin their lives. To reclaim well-being, AA contends that they have to get to the root of the problem, which is not necessarily their drinking but their inability to deal with their feelings in ways that aren't destructive.

AA is able to save people's lives in large part because it has fully embraced the power of culture and its essential mechanisms to help people shift in more positive directions. But you don't have to be a part of this group to leverage the power of culture to manage your emotions. To pop the hood and see how AA, and other effective cultures, manage such a feat, I find it useful to break culture down to three core components: *beliefs and values, norms,* and *practices.*

Popping the Hood on Culture's Emotional Engine

I grew up in a pugnacious corner of Brooklyn, with its own set of deeply entrenched values, and one of my earliest memories is getting punched in the nose by a burly preschooler nicknamed Big Rich. I had made the mistake of refusing to give up control of a highly coveted train set, which didn't fly with Big Rich (who was, I realize now, in fact not that big). When I went home that day and told my parents, my mother responded by kissing my cheek, looking me in the eye, and saying, "Next time you see him, you punch him back!" My father said much the same when it was his turn to counsel me, reflecting the high degree of cultural attunement between my parents on this issue. Needless to say, I got the message. The next day at school I marched directly up to Big Rich and walloped him. In retrospect, not my finest moment, but I was only four. And I had just been taught a very clear cultural value: We don't back down, we fight back.

Beliefs and values are what the people in each group care about, and they can vary wildly from culture to culture. Case in point: Whereas my culture of origin taught me to meet force with force, if you were to travel straight north, to the Arctic Circle, you'd find a group of people who teach their kids to do the exact opposite, *Never fight back.*

In 1970, the anthropologist Jean Briggs left an indelible

mark on the scientific world when she published her account of living for eighteen months in a community of Inuit people called the Utkuhikhalingmiut, or Utku for short. One aspect of life there she observed quite closely was the interaction between the Utku and the white men who came to the mouth of the Black River every year to fish. They usually asked to borrow one of the community's two canoes, and the Utku graciously lent it. But during the time Jean was there on her research trip, living with a host family, the fishermen had brought back the borrowed canoe with a huge hole in its canvas shell and then callously (thought Jean) asked to borrow the remaining one. Watching her host father agree, without protest, to lend out the entire community's last canoe, Jean felt a flash of anger and spoke up on behalf of the Utku: She confronted the leader of the white men and told him in no uncertain terms that he was being selfish, thoughtless, and endangering the community's ability to fish and survive.

She won the argument—the leader quickly backed down, surprised yet not too perturbed—but at a cost: She was spurned by the entire Utku village for her outburst (and her adopted father still lent them the canoe). And she realized something she should already have known: that *anger* was not okay there. Fighting back was never okay.

In Jean's own culture (she was born in America) expressing flashes of anger and frustration was no big deal. In the Utku culture, it was decidedly taboo. They of course got angry like anyone else, but they rarely *expressed* it. Why? One interpretation is that over thousands of years, the Inuit culture had evolved to help them survive in an extreme climate that was just barely conducive to human life. They lived most of the year in close quarters; they had to collaborate intensely to gather and share resources; and they had to cope with hardship and isolation. And so, their culture developed and nurtured a values system that allowed them to thrive within those constraints—one where social harmony was valued above everything . . . even the last canoe.

When I read this account, the Utku's approach to conflict and anger was completely foreign to me; that kind of behavior wouldn't get you far in 1980s Brooklyn. Meanwhile, I'd likely never have survived on the Arctic tundra if I'd gone around popping people in the nose when they upset me. The difference between the cultural values systems, at a fundamental level, could not have been starker. But the two cultures did have something in common: They'd both developed a scaffolding to support their respective values—a way to make their belief systems visible and actionable. We call them *norms.*

Norms are the spoken and unspoken rules that keep us in harmony with others and help promote and maintain a culture's beliefs and values. For instance, a value of our justice system is respect for the law. A norm that upholds that value is the way judges are addressed by "Your Honor" and attorneys are referred to as "counselor" when they are in court. Norms are doubly important because we are a social species. Coordinating our behavior and cementing our beliefs through norms are effective in part because they are linked with something primal—our desire to belong. Norms influence our behavior because we want others to accept us, and to do that, we need guidelines to help us stay within the boundaries of what's okay (we also don't like to get punished, which is often the consequence of violating norms). In this way, norms are an incredible tool for reinforcing the beliefs and values of the group. A little like a sheepdog herding everyone in the same direction.

Norms can be communicated explicitly, like the laws of the Constitution, or we can absorb norms implicitly, through our everyday interactions—kind of like the way you just know it's not okay to bring plastic cups and Styrofoam plates to a party that your environmentalist friends are throwing. These implicit understandings come, in part, from observing what rules others are following. But we also learn by observing what happens when we break norms.

Social exclusion is one tool that cultures around the world use to punish norm violations, and it's as painful as it is ancient. Hundreds of thousands of years ago, ostracism was life threatening. Without the protection and support of the group, you were left vulnerable and exposed. Research shows that ostracism and the social pain that it accompanies are among the most painful emotional experiences we endure as humans.

Breaking the rules = social *exclusion* = very bad.

Following the rules = social *inclusion* = very good.

This provides a plausible explanation for why going against social norms and not fitting in causes emotional and physical distress. Back in the day, conformity wasn't a choice, it was a necessity, and that internal wiring is still with us. But cultures consist of a third element to help us follow those unspoken rules and gel more successfully with others in the group.

Practices are behaviors—like rituals, exercises, and teachings—that help the group actualize their beliefs and successfully adopt social norms. In Hollis's case, when she sat down and started making phone calls, she was leaning on a well-established practice of AA culture. The norm is supporting one another through moments of crisis and possible relapse; the practice is simply *picking up the phone*. Other cultural practices might include prayer, ceremonies, and rituals that further reinforce a culture's values and norms. Which bring us to another arena of our lives that uses this same cultural blueprint— *religion*.

Religion is "the big kahuna" of culture. More than 84 percent of people on the planet are members of religious groups, and research shows that compared with nonreligious folks, people who are religious enjoy better cardiovascular health, less depression and anxiety, and a greater sense that their lives have meaning. Not everyone, of course, joins a religious community or practices their religious beliefs with the intention of regulating their emotions, but it is a powerful ancillary benefit.

Like all cultures, we can think of religion as a recipe consisting of multiple ingredients that help us manage our emotions. The first one is *belief:* Many religions teach that emotions are within our control (for example, through practices such as prayer or meditation) or that a higher power will take care of us (that is, God will provide). The belief in that higher, transcendent power helps people tap into a sense of both awe and connection. This gives people a broader perspective and reminds them that they aren't necessarily the center of the story. Acceptance, gratitude, love, and compassion are central religious experiences that are encouraged and sought by people of many religions. These emotional experiences allow people to reduce their focus on the "self," helping us feel more connected and less immersed in our problems.

Practices such as fasting, ceremonial cleansing, pilgrimages, and religious holidays reinforce beliefs and provide us with a sense of control during times when control is in short supply.

This is likely why so many people turn to rituals when terrible, unfair things happen (think: war, terminal diagnoses, and so on). Cultivating a sense of calm, control, and familiarity, even on a small level, can help us grapple with big emotions. For a practicing Catholic, ducking into a cathedral even thousands of miles from home might feel like taking a deep breath. For others it might be a church picnic or Shabbat dinner that feels easy to join because you already know "the rules." In this way, religious culture is helpful in times of need, which is possibly why there are more people who are religious than there are people on social media (if you can believe it!).

Even if you're not religious, there are still plenty of ways for you to benefit from some of the active ingredients that underlie religion's emotion regulation benefits. One example: Create your own rituals.

A ritual is a sequence of behaviors that are infused with meaning that you do the same way each time. Performing a ritual provides people with a sense of order and predictability. It's called compensatory control: The control you exert over your immediate situation—this ritual you're performing, with its prescribed steps—compensates for the bad feeling you might have in a situation that feels *out* of control. And while you can of course perform a ritual by yourself, it's also beneficial to do them with others because it gives you a sense of solidarity.

The NBA superstar LeBron James is just one of many professional athletes to lean into the power of ritual. Before many games, on the sidelines, he pours chalk in his hands, tosses it into the air, opens his arms wide, and looks to the sky. Afterward, he claps his hands and then blows into each fist a couple of times while the crowd, who by now is in on the ritual, goes wild. For someone about to jog onto the basketball court with a huge amount of performance pressure on his shoulders, it's easy to imagine how this short practice would help distract him, provide him with a sense of control, serve as a way to connect with the fans, and prepare him for what comes next.

Figuring out which ritual will help you in your quest for emotion regulation means looking at your life and identifying where you could use some help. If your anxiety usually spikes around bedtime, there are lots of options, from lighting a candle, to reading your favorite poem, to journaling. Meditation is a particularly useful practice that is no longer necessarily tied to specific religious practices and used by millions as a personal ritual that invites calm. Other rituals could involve the senses so you're using two levers in one: think predinner dance party to get everyone off their devices and to the table, or a morning walk without headphones so you can shift your attention away from your stress-provoking to-do list and toward the newly budding trees and chunky-cheeked squirrels. Rituals, like other practices, can help you live your values and uphold norms. They also act as social glue when practiced as a group, binding you to other members of your culture and strengthening those connections.

Ultimately, what makes AA's culture so effective for emotion regulation isn't unique to AA; it's the fact that it uses *all* the levers of culture to help people. For instance, AA is a culture that deeply *values* personal accountability and acceptance. As a result, there are *norms* (in AA they call them "suggestions") such as staying out of romantic relationships until you are well into your recovery and treating others in the group with patience and understanding.

There are also *practices* in AA that help people adhere to these norms and live their values. One practice involves pairing newcomers with "sponsors" to help them stay on track with their recovery by gently reminding them of the program's "steps" and suggestions. Another involves taking inventory of difficult emotions and resentments that might derail recovery. Many AA members like Hollis credit the group's emphasis on fellowship as the most influential part of the culture. People who have been in the program for years go out of their way to make sure newly sober members are taken care of by offering guidance, phone numbers, and hugs. One of their most closely held tenets is "let us love you until you can love yourself."

• Beliefs, norms, and values exist in every culture; looking specifically for those three components—and how they're working together—can help you find the cultures in your life that are most supportive. And if there is a particular group you really value, but you realize is falling short, you can use this cultural blueprint to figure out where the group's values, norms, and practices could be made more explicit or robust. If you have a group of close friends from college, for example, and a deep value of that group is being there for one another, but you don't have any solid norms and practices that support that—for instance, a norm of hanging out once a month, and a practice of taking turns picking the activity and organizing—that culture might start to feel brittle. The positive cultures of the groups we belong to need nurturing; they won't necessarily persist on their own.

Meanwhile, some cultures in our lives may be doing more harm than good when it comes to our emotional lives. It might be that a foundational belief of that culture is something we don't share—a friend group that does a lot of gossip and backstabbing could be one example, or a work culture that values overwork and busyness. It might be that the beliefs and values that are stated sound great, but the norms and practices that cultivate a positive environment conducive to emotion regulation are not present. As much as this cultural blueprint helps you identify the cultures in your life that are working, it's also a tool for diagnosing the ones that aren't.

Sometimes, one of the cultures in your life might be enough of a negative force on your well-being that it's worth considering whether this is a culture you even want to be in. People opt out of toxic cultures every day. They quit jobs. They change careers. They leave bad relationships. Often for very good reason. If you find yourself in a culture where you cannot live the kind of emotional life you want and need to live, you have two choices. One, move away from the culture. You *can* leave cultures; we do it all the time. It's not always pleasant to extract ourselves from a group that we've been a part of, but sometimes it is the best possible option.

Another option: Change it.

Agent of Change

In 2017, England's famed Football Association needed to shake things up after a series of disappointing tournament losses by the high-profile men's national team. To change what they saw as a problematic culture among their teams, they began by looking for a person to shift the culture.

Their goal was to create "a high-performance culture and psychological resilience across all men's and women's national teams." For this task they found a worthy candidate in Pippa Grange, a seasoned psychologist who had previously successfully improved the culture of several sports teams, organizations, and businesses internationally. Despite the odds stacked against them, and a team history of choking during penalty kicks, England made it to the men's World Cup quarterfinals in 2018, the first time since 1990. After the team won a historic penalty shoot-out earlier in the tournament, the *Daily Mail* began calling Pippa Grange the "Mary Poppins of football" and singing her praises. How did she do it? That's what all of England wanted to know.

It turns out to have involved tapping into those very same levers of culture that we just talked about.

When we think about culture, it feels amorphous, which can be intimidating. Especially if you want to change it. But that's where the blueprint of beliefs, norms, and practices comes in, because it shows you how you can strategically pull those levers and make change. These fundamental tenets of culture that are great to lean on when a culture is working are the exact same places to zero in on when a culture needs to shift. So let's look at how Pippa accomplished this turnaround with the men's national team, along with some other examples.

Rule 1—Check the foundation. If there is something happening within the group that you don't like, look at the beliefs and values that are driving it.

That is exactly what the seasoned leader Dara Khosrowshahi did when he was hired to helm a floundering company in 2017. Far from an easy job, Dara was tasked with cleaning up a number of crises. There was an FBI investigation, splashy headlines about a toxic work culture rife with sexual harassment, and the messy ouster of the company's infamous founder. What the new executive found was that the values that had underpinned the start-up's meteoric rise were now contributing to its downfall. Describing the culture under the previous CEO as "unrestrained," *The New York Times* reported on a workplace environment where people were physically and sexually harassed, threatened with violence, and subject to "homophobic slurs"—an emotionally damaging culture, to say the least.

Of course, this was no ordinary company. Founded in 2009, Uber was being hailed as the next great disrupter of Silicon Valley, and by 2015 it was the highest-valued start-up in the world at fifty-five billion dollars. Creating a more emotionally healthy culture while maintaining their success was the needle that Dara had to thread. An important part of this culture change process that Uber's new leadership team got right was to establish concrete *values* instead of ones that were vague or abstract. Two of the new values they eventually settled on were "build with heart" and "stand for safety." On their values web page they drill down into the specifics of what these beliefs mean—for instance, actively caring about their customers and becoming the most vigilant company in the business when it comes to safety. Statements like these create alignment within the group and also spell out what otherwise goes unsaid, making values visible.

Rule 2—Rethink norms. In 2012, Google did a study to figure out what made its teams succeed or fail. Because it is a company famously obsessed with maximizing productivity and researching group dynamics in the service of innovation, it embarked on a massive data-crunching operation to better understand the elements that contribute to teams thriving. They began thinking that it had something to do with *who* was in the group.

But they were wrong.

What they found was that a key igniting agent for these groups was the *norm* surrounding social interaction. Google found that its most successful groups were governed by norms that promoted psychological safety—giving one another time to talk, being more aware of the feelings of others, and sharing difficult moments and feelings with the group. Google took this information and applied it to improve the communication within all its teams. For instance, it emphasized the need to embrace disagreements, welcome seemingly "silly" questions, and be open about failure.

Not every company, group, or organization has the financial heft and data chops to do a full-blown research project on which norms are best. But they do have the capacity to do a kind of norms audit, where you identify what the norms are currently and cross-check them with the group's beliefs and values. Are they in service of those goals? Or could some unwritten rules be changed to better fit what the group needs? The good news is norm changes aren't as hard as you might think. Most people want to follow the norms of their group; we want to belong and fit in. Because of this, people are already inclined to follow the rules; you just have to communicate them effectively.

Rule 3—Customize your practices. When it comes to practices, there are many opportunities to create tools for helping people adjust to new cultural beliefs and norms. This is a bespoke matter, but really it boils down to asking what would help people cope with their emotions within this culture while staying aligned with the values and norms of the group. *Practices* were, more than anything else, at the heart of the culture change within the men's national team under Pippa Grange. When Pippa came on board, she and the men's team coach first discussed new values and norms. They wanted to improve players' resilience and bet that getting there required helping them build a sense of emotional safety. In one interview with *HR* magazine, Pippa emphasized the importance of building "care and intimacy" within the team, and said, "If you know people are in the trenches with you then there's a sense of safety in that—a sense of home and belonging."

One new value they identified to support that goal was to truly privilege trust and understanding of one another as teammates, and a norm they established around that value was the expectation that players share their feelings, support one another, and encourage vulnerability. But to actually make this possible, they had to create specific practices to help people get there.

What they came up with was a series of small group conversations led by Pippa. In these intimate moments, the men would talk about their fears and insecurities, embracing one another's concerns instead of adhering to the stiff-upper-lip status quo. One player who was going through a severe depression went public with his struggles, a risk he likely wouldn't have taken before. Apart from these heartfelt conversations, Pippa encouraged the players to have fun with one another, practicing play, if you will. The group played games (other than football) together that built coordination and trust, but also filled the players with more lightness and joy. After their big win against Tunisia in the 2018 World Cup, the staff filled an indoor swimming pool with a herd of inflatable unicorns.

I wish I'd been a fly on the wall that day, to see football players racing one another on purple and white floaties.

What Pippa Grange understood was that you can't just lay out new values and norms and expect people to immediately start embracing them. You've got to help them "train" for it. The practices that she and the team established were crucial for that purpose. Instead of just saying "be vulnerable," you have to create a psychologically safe space in which to do so, model ways to do that, and embrace practices that build the ancillary skills needed to act in accordance with the new values and norms. One of the most important lessons we can learn from Pippa Grange is that changing culture for better emotion regulation doesn't have to be boring. In fact, you can make it fun.

Accessing the Cultural Control Panel

Much like the air, or the water, or the sunshine, culture is around us all the time, and as a result sometimes it can be hard to see the ways that the groups we're a part of affect our emotional lives. This is in part why culture clashes can be so difficult to manage: because we often aren't even aware of why we, or the groups we're interacting with, are thinking, feeling, and behaving the way we do. But by understanding the way culture shapes our emotional lives we have a blueprint for how to harness this shifter better. It's like knowing which buttons on your smart switch control the blinds and which ones control the thermostat.

This brings us to a final question: Why do we so often find ourselves in situations where that master switch is right there, but we *just don't press it*? This is a classic conundrum of the human condition: We can understand exactly what we need to do and still fail to do it. How do we bridge the gap between *knowing* and *doing*?

Before you started reading this book, you no doubt were aware of a grab bag of emotion regulation tools. And yet, how many times have you successfully used them in the heat of the moment to shift out of an emotional funk? How often have you preemptively planned for how to use these tools in your day-to-day life? Maybe your friend persuades you to go for a walk one day after a bad breakup, and you feel better, but will you do it on your own the next time something tough happens? Maybe you know that 1990s hip-hop always makes you feel better, but do you have a playlist all queued up on your streaming app?

Most of us struggle with opening the toolbox in the first place because even throwing back the lid can feel like a Herculean task when we are in distress. *We know, but we don't always do.* Why do we keep failing to do what we know and what science teaches us will make us feel better? And what can we do to fix this?

To answer these questions, let's revisit an old friend.

Part Four

Shifting by Design

Chapter Nine

From Knowing to Doing: Making Shifting Automatic

It was 3:00 a.m. when Matt Maasdam, along with ten Navy SEALs, stealthily approached a house in a middle-class neighborhood outside Baghdad, Iraq. It was hot and dusty even at that time of night, with most residents either sleeping together in the coolest interior rooms of their homes or bunking on the roofs to catch a breeze.

The SEALs didn't know much about the household where the bomb maker they were searching for was living, other than that he was there. They didn't know how well armed he was, or how many innocent civilians were slumbering inside. But they knew that bomb makers were booby-trap experts. They often buried explosives along the perimeter of buildings where soldiers would line up, hoping to kill a dozen men with one blast. They also knew that bomb makers' homes were heavily guarded, with armed men ready to pick you off from second-story hiding spots, and that it was possible they'd find insurgents napping in suicide vests, trip wires hooked to projectiles, or women with grenades in the pockets of their burkas. There were plenty of unknowns, plenty of ways that the mission could go sideways and get them all killed. But Matt's team was ready.

The Navy SEALs are one of the most elite organizations in the world for many reasons, and a critical part of their success is the meticulous planning they do before missions. Before a mission, SEAL teams carefully prepare for every contingency they can imagine. During the days leading up to their operation, Matt's team walked through more than a dozen potential scenarios.

If the helicopter they were traveling in loses power . . .

If the team takes fire on the way to the target . . .

If there are five times more insurgents than they anticipated . . .

If someone steps on a land mine on the way out . . .

The mission leaders would drill each scenario into their heads by calling on them Socratic style.

"Matt, if _____ happens, then what's your next move?"

For every "if," there were three "then's"—specific plans for responding automatically when things go wrong, and getting the help they need to carry out the mission and still get out alive.

So, when Matt and his team approached the bomb maker's house with their night vision goggles and quiet breath, they had an initial plan. They weren't going to line up against the perimeter of the house. They would scatter themselves in the front yard, and as soon as the charge on the front door blew, they would move rapidly in a coordinated fashion, like a basketball team running a fast-break play. Half of the SEALs would run through the door, and half would flood into the rest of the house through other doors and windows.

When they blew the charge, they encountered a contingency they *didn't* anticipate. The exterior walls of the house were mostly glass, and as soon as the charge blew, they shattered. From the street, it looked like the house exploded, with giant shards raining down on the SEALs from twenty feet above. Remarkably, no one got hurt. In training they had practiced blowing up all kinds of structures and responding to the myriad ways different kinds of materials collapse. As a result, Matt's team wasn't overwhelmed or taken off guard when the shards of glass came raining down, even if they were a little surprised. They simply "turned steel," letting their armor take the blow as they crouched and covered, allowing the glass to impact their body armor before continuing with the mission. While the glass explosion certainly woke up the neighborhood and heightened the need to move swiftly, the SEALs had been well trained to go about their business despite the unexpected. In the end they were able to go in, apprehend the bomb maker, and get out safely.

When Navy SEALs want to learn how to do something, they find the best people in the world to train them. When Matt was still on active duty, his team found that it took them ten minutes to change a flat tire on the battlefield, which felt perilously long to be stationary and exposed. Because seconds can make the difference between life and death, the SEALs reached out to a NASCAR pit crew and spent two days learning how to change tires more efficiently and practicing until their fingers were numb. After those lessons, they went from being able to change a tire in ten minutes to well under two minutes.

This is *exactly* what we want to be able to do with our emotions. We want to be like a NASCAR pit crew when we need to shift. Whether we want to shift *into* a more positive state or *out* of a painful one, the goal is to be able to do it more smoothly, intentionally, and swiftly. And while some of that has to do with knowing what tools we can use in the moment—the tools we've gone through in this book—we also need to know how to *implement* skills swiftly and efficiently when we need them. Which raises the question: How do you go from *knowing* to *doing*?

The Missing Link

In the late 1980s, Gabriele Oettingen, a German-born psychologist, came to the United States to study a topic that has been fascinating people for ages: daydreams. She was interested in fantasies. Those far-flung simulations about what could be in the future that all of us engage in from time to time. Was there any functionality to them? Did jumping into the mental time travel machine to contemplate the improbable help bring those goals to fruition? Going into the work, she thought it would.

She was wrong.

The idea that we should visualize positive outcomes to attain success is well documented. But Gabriele's studies showed that not only did it *not* predict success; it predicted the opposite. In study after study, she found that cuing people to fantasize about the things they wanted actually *decreased* the likelihood that they'd achieve their goals: When they imagined the future, they got a little satisfying taste of that wonderful feeling of having achieved their desire, and it undermined their effort toward it. They were, essentially, lulled into feeling that they'd already accomplished what they desired.

Now, fixating on the obstacles in their way wasn't helpful either. Gabriele called this "dwelling"—focusing so much on the blockers in your way, that you can't see past them to the future you so desire.

When it came to fantasy, both dwelling and indulging were bad for you on their own. But here's the crucial piece: They were great *if you linked them together.*

If you indulge in visualizing what you want, but never look at what you have to overcome in order to get there, it's worse for you. And if you're dwelling in the weeds and only seeing obstacles, you can't see a way forward or muster up the energy to get there. Gabriele's major discovery was that if you both vividly fantasize about the future you want *and* fully visualize the problems you're going to run into on the way there, outcomes shoot way up, in terms of both goal accomplishment and emotional wellness.

She broke down successful goal pursuit into three distinct steps:

Step 1: You visualize your *WISH*—what specifically you desire. *I want to finish writing this book!*

Step 2: You visualize the OUTCOME—the best thing about achieving that wish, and how it will make you feel. *The book will be published, and I'll feel a huge sense of accomplishment about getting this guide out into readers' hands.*

Step 3: You visualize the OBSTACLE—the main challenge *within yourself. I say "yes" to so many other asks from people who want help that I don't save enough time to write and never finish the manuscript.*

WISH.

OUTCOME.

OBSTACLE.

Gabriele called the three-part plan "mental contrasting" because what this involves is identifying your "wish" and then *contrasting* the outcome you hope to achieve with the crucial obstacle that stands in the way. This was a significant step forward for figuring out how to help people convert their dreams into reality, but there was one crucial missing link. Now that I've visualized my wish to write a book, imagined how I'll feel when I do, and then identified the potential obstacle, what do I *do* to overcome this roadblock?

Luckily, another researcher, Peter Gollwitzer, completely independently, happened to be simultaneously working on that missing piece: the psychology of *planning*. Coincidentally, Peter happened to be Gabriele's husband.

Around the same time that Gabriele was studying mental contrasting, Peter was digging into how people approach making *plans,* and what factors might increase the likelihood that they follow through with them. We make all kinds of well-intentioned plans—*I'm going to eat better. Lose weight. I'm going to save money, be a better parent, stop drinking, talk to my dad about that thing that's been bothering me for a decade, be less anxious, not blow up at my kids so much*—and then fail to see them through and feel terrible about it. There are countless reasons: life gets in the way, we get overwhelmed, we get distracted, we get stressed, we get scared. Peter wanted to know how we could make plans stickier. So, he came up with a remarkably simple and effective solution. He called them "implementation intentions."

Peter's intuition was that you had to make it *easier* for people to follow through with their plans when life gets in the way. The goal was to make the process less effortful. What he found was that you could do that by using a framework similar to what the SEALs do when they prepare for a mission: create an if-then plan.

IF the alarm goes off and it's 6:00 a.m., THEN I'll grab my duffel bag and go to the gym.

IF I get hungry before lunch, THEN I will have one of the apples on the counter.

IF I feel anxious about the week on Sunday night, THEN I will make a to-do list, put it away until Monday morning, and listen to a feel-good song.

IF I get overwhelmed and angry during the meeting again, THEN I will pause and think about the fact that we're all on the same team before I decide what to say.

The beauty of this approach is not only that it forces people to think through high emotion scenarios ahead of time to envision how they'd act. It also makes exercising goal pursuit—in our case, the goal of managing our emotions effectively—as automatic as a handshake.

That's our goal with emotion regulation: You don't have to think about what to do when the situation presents itself. You just do it. And that can be a tremendous boon for managing our emotional life, because remember, we follow the "law of least effort." We don't like to do things that are hard. And it can be hard to think about what strategy to use in a moment when you're bathed in emotion.

Implementation intentions take the effort out of the formula. They forge a link in your mind that you can rely on when a situation overwhelms you. They work on two levels: they connect your goal with the specifics of how to reach that goal, and they bookmark the situation with context clues that make it more likely you'll be able to remember what to do. When we are rehearsing these situations in our minds, we are creating psychological links between specific situations and specific actions. And that makes it more likely that we'll convert our knowledge into action.

Gabriele and Peter spent decades independently trying to crack the puzzles that got them interested in psychology. Then, around 2010, a lightbulb went off. They realized they could put their steps together, and they were better than the sum of their parts.

Enter the tool called WOOP: Wish, Outcome, Obstacle, Plan.

WOOP There It Is

WOOP is the marriage of two independently studied ideas that emerged from a literal marriage. And it helps us take the knowledge we have acquired about how to manage our emotions and put that knowledge to use in our lives.

The "mental contrasting" piece (WOO) helps energize people around their goals and specify the obstacle in the way. The "implementation intentions" piece (P) fuses each obstacle (the "if") to a specific action (the "then") and makes the entire enterprise of regulating our feelings more effortless.

So how do we use WOOP with emotion regulation? Here are a couple of examples.

Wish: I want to be more patient with my children when they irritate me (yes, Maya and Dani, it happens).

Outcome: I'm going to have a better relationship with them and be a better father.

Obstacle: When they call each other stupid, I sometimes lose my temper—I grew up in an atmosphere where people put each other down, and I'm really reactive to that.

Plan: Now I'm going to build a plan around my obstacle.

IF they are fighting, THEN I'm going to zoom out, remind myself that they're kids, my wife and I acted similarly when we were young, their brains are still developing, and then get their attention without yelling.

This is the sort of circumstance we might encounter on a daily basis. Now I want to take you through how you might use WOOP in an extraordinarily difficult emotional situation, because at some point or another in life, we all find ourselves in one of those.

I heard a story recently that stuck with me. A friend of

a friend lost someone close to him very suddenly: a younger brother died by suicide. It was a shock to the whole family. A shattering, immeasurable loss. In the months after, he was reeling, looping in and out of intense grief. He was managing—going to work, going to therapy, processing. But he found that sometimes the sadness would well up in the worst ways. He has young kids and noticed that it was affecting his time with them. They'd be at the playground, playing monster-tag (he was the monster), and suddenly he'd get hit with the loss, and he'd have to walk away. He didn't want to run from his grief. He understood he needed to mourn his brother. But he didn't want to miss his kids' childhood while doing so. How can we WOOP such a deep loss? Here's what one version looks like:

Wish: "To be able to hang out with my kids and enjoy my life, even during this era of intense grief."

Outcome: "When my kids are grown up, they'll still look back on this time and remember a lot of joy."

Obstacle: "When the grief hits, I can't see my way out of it. I completely lose perspective."

Plan: "IF I'm with the kids and I start to feel it, THEN I'm going to stop and focus on the future. I'm going to take ten seconds and imagine them as grown-ups, and me looking back on the whole sweep of my life. I'm going to remind myself that in the grand scheme of things this is just one short season."

Try using WOOP for an emotional challenge you're facing. Choose something and try it, right now.

W = Wish (*Write a wish that is important to you—challenging but feasible.*)

O = Outcome (*How will it feel when you accomplish this?*)

O = Obstacle (*What is the personal obstacle?*)

P = Plan (*What's the action you're going to take when faced with this obstacle?*)

So fill in: if ____*(obstacle)*____, then ____*(action)*____.

The goal is to be able to shift your emotions easily and effortlessly—almost habitually, the way you buckle your seat belt without even thinking about it when you get in the car. If that sounds impossible, just remember the Navy SEALs: What they do isn't easy, but with enough planning and practice, it can become close to automatic.

Over the past two decades, several studies have shown the power of WOOP and its long-term, lasting effects on people's lives. Using WOOP leads to better success for students with studying and grades, better working through of negative feelings, healthier eating and exercise behaviors, people with depression taking better care of themselves, and more thriving relationships, among others. And, crucially, this "technology" is *easy to use*.

In the mid 2010s, a group of German scientists took this into the classroom with first graders in the context of a large field experiment. They talked about WOOP, guided the kids through it, and did it with them five times. What they saw was remarkable: These little six-year-olds managed their emotions better and earned better grades. Moreover, the effect that WOOP had on their lives was long lasting; the experimenters found positive effects of the intervention with the kids three years after its implementation.

Of course, we can't plan for everything. As much as you can visualize your desired emotional futures and plan out if-thens, you can't always predict what's going to happen. When the SEALs went into raid the bomb maker's home, they'd run if-then scenarios over and over, but they hadn't trained for the specific contingency they encountered when the glass shattered into thousands of razor-sharp shards. Still, they knew how to respond when the charge went off: They had a set of tools they'd used in similar circumstances, and they automatically reached for them.

The bombs most of us will face are metaphorical, not literal. But the point is that unexpected emotional challenges *will* come along, and you may be deep in the emotional soup before you realize it. And when this happens, you have tools to fall back on in these situations: the shifters you've learned about in this book.

Beyond Silver Bullets

A critical piece to remember: *There are no one-size-fits-all solutions*. Like exercise, nutrition, and medicine, emotion regulation is highly idiosyncratic. Back in chapter 3, I mentioned a study I collaborated on in 2020, during the onset of the COVID pandemic, that was designed to reveal how people were coping with the anxiety that had swept the globe. We wanted to know whether a particular tool, or combination of tools, was emerging as somewhat of a universal regulation method during this incredibly challenging time.

What we found: There was *wild* diversity across participants in the tools that benefited them most. Some people used more than a dozen different tools in a day; others used none. What worked for one person had no effect for the next. And for many people, what helped them varied from day to day. All this is to say, I encourage you to open up to playing around with these shifters. Find the tools that work for you, and once you do, work on strengthening the likelihood that you will *use* them when you need them the most via WOOP.

Of course, no matter how much we learn and practice and shift, some emotional experiences will always be difficult. If emotions were a cinch to rein in, they wouldn't be useful, because we'd never permit ourselves to feel bad. Just like with physical pain, there is a reason we experience emotional pain: It's information that we need.

The goal is not to eradicate emotional pain in our lives or make every instance of conflict feel like rainbows and sunshine. The goal is to listen to our emotions and respond to them in a healthy way. It's when we're stuck in a painful emotion that we might need some extra help shifting. Turning the volume on our emotions up or down is often a matter of degrees. But degrees can matter a great deal to the cumulative emotional experiences of our lives and to the lives of those around us. And in moments of great distress and confusion, the skills we have developed are there to catch us like a safety net under a perilous tightrope, a lesson I've learned myself a few times when I've slipped, fallen, and then found—to my surprise—that the net was there.

It's 5:00 a.m., Do You Know Where Your Emotions Are?

"School will be canceled for today. Please see the email message for more information."

That's the text message I received at 5:07 a.m., about a week before I sat down to write this conclusion. Bleary-eyed and nearly knocking over a water glass as I groped around, I picked up my phone and saw that the source of the noise was my daughter's school. With my wife still asleep beside me, I tossed the phone aside and tried to get back to sleep. But after a few minutes of lying there, I realized something was nagging at me. The text message didn't say *why* school was canceled. Whenever there is a snow day, the school signals that it's "due to inclement weather." But having repeatedly checked the weather during the previous day (an occupational hazard of living in Michigan with school-age children during winter), I wasn't expecting a storm. So, why were they canceling school?

I picked up my phone again and checked my email this time. They were canceling school because of two anonymous email threats that were sent to staff and students late that night.

What the hell? What kind of a threat?

I felt my heart begin to pound, and my mind fired up almost immediately. I knew from experience there would be no getting back to sleep, so I got up and made tea, all the while rechecking my email for more information. The school was short on details but did say that they were following up with police to investigate. Of course, the questions abounded.

Was the threat credible? What type of threat was it? Bomb threat? Guns? Could they trace the email? Would they expect us to send our kids back to school even if they couldn't figure out who sent the threat?

Luckily—or not, depending on your perspective—another anxious parent was up early too and forwarded the threatening email to me, which she had received from one of the students who was copied on the original message. It said, "I'm coming to your school tomorrow and shooting it up. . . . After that I will commit suicide. Just a fair warning for you guys. I'm considering also bringing a bomb too. Have fun rotting."

These are words no parent or educator wants to read. I was no exception.

When I was researching my first book, *Chatter,* I wrote about shooting massacres at Northern Illinois University and Virginia Tech, which probably sensitized me even more to the specter of such a tragedy so close to home. As I stood there in the predawn light, horrible images that I had seen while researching those stories wrenched themselves from my mental archives and floated to the surface. This knowledge, combined with the prevalence of mass shootings in schools across the

United States, left me shaken. Then of course there was the fact that my fears weren't actually unreasonable. Just two years ago, there had been a deadly shooting about an hour down the road from us at Oxford High School in Oakland County, Michigan. A distance that felt much too close for comfort. *Could it happen here?* seemed like an all-too-real question.

I fell back immediately on my emotion regulation tools. I zoomed out and reframed the situation. Though this was scary, no one was hurt, and there was nothing I could do about what might happen with this situation anyway. And I deliberately refocused my attention. Exercising some healthy distraction, I leaned hard into my work. By the time my wife came down for tea (a coffee-drinking family we're not), I was already engrossed in work on this book.

Pretty quickly, the FBI was called in to investigate, and the school shut down while they traced the source. In the two days that school officials and law enforcement were sorting things out, there was still a looming uncertainty about whether they'd find the person responsible. During this time, it would have been easy to get caught up in the collective chatter that had spread throughout our community. Digital social contagion was on full display as parents and students took to text messaging threads to vent their fears and frustrations. But I largely resisted the temptation to engage in the online venting sessions (remember, a *little* emotional sharing and validation is healthy). I strove to get perspective on the situation, talking to a friend who formerly ran a security firm. I tried to care for others, reassuring my daughter and wife and trying to help them broaden their own perspectives. And finally, when I sensed chatter setting in, I took a sensory break and changed my space by going for a walk.

All these actions helped me turn down the volume on my fear. I was by no means devoid of emotion, but I wasn't overtaken by it. Remember, the goal is never to turn off fear altogether when it's triggered in the appropriate circumstances. That's a functional response that keeps us vigilant. Each of the small moves I had made pushed me ever so slightly, and sometimes greatly, in the right direction. And as I successfully shifted my own emotions, I noticed that my daughter's emotions seemed to be more regulated as well. Likely because kids are sponges when it comes to their parents. They catch our feelings.

I'm not telling you this story to highlight how naturally gifted I am at shifting. Far from it. There have been plenty of times when I've failed—miserably—to manage my emotions (if you're curious about what one of those looked like, just read the introduction to *Chatter*). I'm human. A decade ago, before I fully understood the dangers of co-rumination, I would have dived headfirst into the group text venting sessions and stayed there longer than I should have, ramping up my own anxiety in the process. I was able to avoid that pitfall and many others because I know what emotion regulation tools are available, how to use them, and which ones work for me.

Learning to play your emotional Stradivarius doesn't turn school-shooting threats into birthday parties. It turns them into experiences that can be managed, and that's the challenge we've been facing for likely as long as we've been roaming the planet—to understand how to skillfully amplify the feelings we want and mitigate the feelings that are causing us pain. Looking back, I realize that is exactly what my grandmother was trying to teach me all those years ago when she told me to not ask "why."

The Nazis didn't get my grandmother, but a second bout of breast cancer at the age of ninety-two did. I was there with her at the very end, telephonically connected. My mom called me from my grandmother's apartment in Miami Beach during the final minutes. I could hear my grandmother moaning in the background, and my mother's weeping front and center. While I wished I could have been there in person to hold her hand in her final minutes, I did get to say goodbye and whisper to her how much I loved her.

In the immediate aftermath of her passing, thoughts about her life and our times together swirled in my head. My grandmother had achieved the extraordinary. She had everything stripped away from her, lived for years in destitution, suffered unimaginable atrocities, and survived. But she didn't merely endure; you could argue that she thrived.

After the war she and my grandfather started a life with nothing, sharing a house in Brooklyn with my mom (an eight-year-old at the time) and my grandmother's sister and brother-in-law who also survived the war. From there it was the classic immigrant story. They worked hard as tailors and earned enough to eventually purchase a beautiful two-family home on a corner lot of a tree-lined street in Brooklyn. They'd go for walks each day in the park across the street and quickly built a community of friends to support them. Theirs was a home full of warmth and noise, blessed by children and eventually grandchildren. A far cry from sleeping in the woods during the winter, scratching her head raw from lice, and evading Nazis.

In retirement, my grandmother transformed into a snow-bird, buying a condo in Miami Beach with the money earned from rental properties and careful saving. She put one of her children through college and graduate school and would eventually help support me in my education as well. As the lucky kid who got to see Bubby every day after school, I know she loved our connection and felt lucky to have a family who relied on one another in such a comforting and familiar way.

Her death wasn't just a loss to mourn. It was a life to celebrate. A life constructed in no small part by her ability to manage the devastating emotions she experienced.

The past was painful, and she dealt with it in ways that I didn't understand at the time. She had in fact developed her own unique brand of coping—not too much self-reflection (at least that I'm aware of), but high doses of family and community support. She also became enamored with landscaping (the Brooklyn version, which meant copious amounts of houseplants), which enhanced her exposure to emotionally fortifying green spaces. She regularly fell back on Jewish cultural traditions such as prayer and Remembrance Day to work through the deeper pain she still carried. Every lipstick-laden kiss she planted on my cheek, every sweet-smelling noodle kugel that came out of the oven, were moving her emotions around in ways that I, as a kid, could not comprehend.

My grandmother's toolbox was totally different from mine. And mine is likely different from yours. That highlights a key point of this book that bears repeating. *There are no one-size-fits-all solutions when it comes to managing our emotions.* And dogmatically searching for them is likely to be as successful as finding a single diet or exercise routine that works for everyone.

The beauty of knowing about the shifters we've talked about is that they give you the opportunity to identify the tools that work best for you. For some that may look like a cocktail of strategies that resemble mine: lots of perspective broadening, a decent dose of healthy sensory regulation, and an exquisitely curated emotional advisory board waiting in the background to be called to action when needed. For others, the shifters they use will take a totally different form. And in some cases, people will benefit from using their go-to shifters in conjunction with additional interventions such as exercise, therapy, and certain medications. You don't have to choose between options. You have to figure out the combinations that work best for you.

This book has given you the shifters for you to use, and it's given you a framework for how to implement them in your life. That's step 1: working with these tools and learning how to shift emotions yourself. The second step? *Share this information with others.*

I began this book with a story about how every time I give a talk or deliver a workshop, I'm approached by people who want to know more about how to manage their emotions. After we chat, many of them inevitably ask me the same question:

Why wasn't I taught this?

I think often about what I was taught in school, and how much (or little) I rely on that information now. And I come back to the same place: my middle and high school classes on the digestive system. There's a lesson that sticks out: learning about the intricacies of peristalsis, the process by which food travels from one opening (your mouth) to another (you know the one). I learned about how my insides were constricting and releasing in a wavelike motion, pushing food through my body throughout the day. The lesson was without question interesting. After all, I still carry this knowledge with me today. But how many times have I used that knowledge in my life?

Exactly twice.

When they were little, each of my two daughters asked me how it was that they could still swallow food while doing handstands. Now ask me how many times I have used my knowledge of emotions—what they are, why we have them, and how to harness them.

Every single day.

Right now, schools vary dramatically in the emphasis they place on formally teaching students about their emotions, a topic that is certainly as central to how they function as their digestive system and arguably just as important for their success in life as their knowledge of math or chemistry. But there are signs on the horizon that the time is ripe for that to change. Increasingly governments and organizations are appointing leaders to spearhead campaigns to promote well-being and combat emotion dysregulation. School leaders are collaborating with scientists to integrate lessons on emotion regulation into their children's and adolescents' curricula. The stigma and shame around sharing our emotions is waning, ceding ground to a generation that is more than happy to acknowledge the central role that emotions play in our lives.

Between eight and ten thousand years ago, we were carving holes in our skulls in a desperate attempt to feel better, and as recently as the middle of the last century we were essentially still doing that in a *slightly* more sophisticated form: the frontal lobotomy. That's an enormous swath of human history where one of our tools for trying to improve people's emotional lives involved making holes in people's heads. We've come a long way since the ice pick days, but we need to do a better job using the information we've accumulated about the science of emotion regulation in our lives. The needle is moving. But awareness still needs to be raised. That means talk to your kids at the dinner table. Normalize these conversations in the workplace—especially if you have a leadership role of any kind. It means supporting public health policies around mental health awareness and removing barriers for people to get support, education, and care.

The world can be unpredictable. We never know what kind of text message we will wake up to. How we handle whatever emotion it elicits shapes everything from the unfolding of a single day in our own individual lives, to our children's emotional world, to our workplaces and communities, to political conflicts that unfold around the globe. It's for these reasons that I genuinely believe that understanding how to manage our emotions is one of the great challenges we face. Although we still have much to learn, we've made enormous strides. And you could argue that one of the big reasons we have come as far as we have is that "why" *isn't* always a crooked letter; it is the beginning of learning that we have the power to shift our lives, and help shift the lives of those we love.

Acknowledgments

Not long after publishing *Chatter* the idea for *Shift* began to percolate. But after coming off a multiyear book project, I wasn't sure I had another one in me. Two people gave me the boost I needed to get back to writing: my partner in life, Lara, and my agent, Doug Abrams. Thank you both for your continued partnership, love, and support.

Shift tells the story of the scientific renaissance that has transformed our understanding of emotion regulation. More colleagues and students then I can name have painstakingly contributed to this work. Readers of *Shift* and I owe you all a debt of gratitude.

Gillian Blake is the model editor. She grasped what *Shift* could be immediately and unrelentingly worked to help me bring that vision to light. Despite being one of the busiest people in publishing, Gillian always found time to respond to drafts brilliantly and incisively (and with empathy and a dose of humor to boot). *Shift* is a better book because of you.

Writing can be a solitary process. But I was fortunate to have a remarkable team support me every step of the way. Lauren Hamlin and Allyssa Knickerbocker, you are my consummate authorial guides. *Shift* wouldn't be what it is without your wisdom, literary prowess, and whit. Thanks, too, to Lara Love, Aaron Shulman, and Rachel Neumann for providing wonderful edits on early drafts; Even Nesterak, Alex Wormley, and Kate Schertz for swooping in to fact-check (again and again and again); and Amy Li and Jess Scott who helped keep the project on track.

Crown, Fortier PR, and the Lavin Agency collaborated to provide a dream team of sales and marketing support. To Allyssa Fortunato, Mary Moates, Mason Eng, Grace Ermias, Charles Yao, Adrianna Stadnyck, Alethea NG, Dyana Messina, Julie Cepler, and the rest of the teams, thank you for your unrelenting commitment.

It's comforting to have other people to chat with when you embark on the herculean task of writing a book. Jamil Zaki and Angela Duckworth were my "book buddies" when I wrote *Chatter*, and they expertly filled that role again with *Shift*. My love and appreciation for you both knows no bounds.

Thank you Dan Pink, Adam Grant, Lisa Damour, Charles Duhigg, Laurie Santos, Maya Shankar, Jason Moser, Micaela Rodriguez, Matt Guttman, and Chayce Baldwin for supporting *Shift* as well as the countless people who shared their stories with me. I think about your generosity often and I am grateful for it.

Thanks too, to the rest of my immediate family—Basil, Irma, Karen, Ian, Lila, and Owen—and my friends (you know who you are) who once again put up with my periodic absences, and the rest of my talented students and colleagues for supporting me throughout the writing process, including the teams at Michigan and Ross Executive Education. There are too many of you to list and I would surely leave someone out if I tried.

And Lara, Maya, and Dani—thank you for being perfect just as you are.

Notes

INTRODUCTION Why *is a Crooked Letter*

xi   **actually a trap:** I relied on three sources to tell my grandmother's story: an oral history that was performed by the U.S. Holocaust Memorial Museum, Dora Kramen Dimitro, interview by Randy Goldman, July 18, 1996, collections.ushmm.org/oh_findingaids/RG-50.030.0372_trs _en.pdf; Yaffa Eliach, *There Once Was a World: 900-Year Chronicle of the Shtetl of Eishyshok* (Boston: Little, Brown, 1999); and memories of my conversations with her growing up.

xii   **used to house animals:** My grandmother's oral account and the written materials about her experience conflicted with respect to who built the *ziemlanka*. I deferred to my grandmother's first-person account.

xv   **Holocaust remembrance day:** This was not the official Holocaust Remembrance Day but rather an event that my grandparents and other co-survivors organized.

xviii   **Until relatively recently:** Daniel Dukes et al., "The Rise of Affectivism," *Nature Human Behaviour* 5, no. 7 (July 1, 2021): 816–20, www.nature.com /articles/s41562-021-01130-8, doi.org/10.1038/s41562-021-01130-8.

xviii   **heartbreak:** Shayla Love, "The Relatable Emotions of Depressed People from 3,000 Years Ago," *Vice*, May 2021.

xviii   **In the mid 1860s, an American diplomat:** Ephraim George Squier, *Peru:*

Incidents of Travel and Exploration in the Land of the Incas (New York: Harper & Brothers, 1877); Hiran R. Fernando and Stanley Finger, "Ephraim George Squier's Peruvian Skull and the Discovery of Cranial Trepanation," in *Trepanation: History, Discovery, Theory*, ed. Robert Arnott et al. (Boca Raton, Fla.: Taylor & Francis Group, 2003), 3–19; Charles Gross, "A Hole in the Head: A Complete History of Trepanation," *The MIT Press Reader* (Aug. 29, 2019), thereader .mitpress.mit.edu/hole-in-the-head-trepanation.

xix   **Dr. Paul Broca:** Students of psychology and medicine will be familiar with the doctor's name. Broca's area, a region of the brain involved in speech, was named after him.

xix   *were still alive*: William T. Clower and Stanley Finger, "Discovering Trepanation: The Contribution of Paul Broca," *Neurosurgery* 49, no. 6 (2001). See also Charles G. Gross, *A Hole in the Head: More Tales of the History of Neuroscience* (Cambridge, Mass.: MIT Press, 2009).

xix   **creation of holes in people's skulls:** M. Ghannaee Arani, E. Fakharian, and F. Sarbandi, "Ancient Legacy of Cranial Surgery," *Archives of Trauma Research* 1, no. 2 (Summer 2012): 72–74; Lydia Kang and Nate Pedersen, *Quackery: A Brief History of the Worst Ways to Cure Everything* (New York: Workman, 2017); Ira Rutkow, *Empire of the Scalpel: The History of Surgery* (New York: Scribner, 2022); Jeffrey A. Lieberman and Ogi Ogas, *Shrinks: The Untold Story of Psychiatry* (London: Weidenfeld & Nicolson, 2016).

xx   **manage their *emotions*:** Miguel A. Farira Jr., "Violence, Mental Illness, and the Brain—a Brief History of Psychosurgery: Part 1—from Trephination to Lobotomy," *Surgical Neurology International* 4, no. 49 (2013). Also see R. Aaron Robison et al., "Surgery of the Mind, Mood, and Conscious State: An Idea in Evolution," *World Neurosurgery* 77, no. 5–6 (2012): 662–86; Rutkow, *Empire of the Scalpel*.

xx   **impossible to know:** Rutkow, *Empire of the Scalpel*, 17.

xx   **trying to find tools to regulate them:** K. Tajima-Pozo et al., "Practicing Exorcism in Schizophrenia," *BMJ Case Reports*, Feb. 15, 2011, bcr1020092350, doi.org/10.1136/bcr.10.2009.2350; Kang and Pedersen, *Quackery;* Ronald J. Comer and Jonathan S. Comer, *Abnormal Psychology* (New York: Macmillan, 2018); Stephen A. Diamond, "Possession, Exorcism, and Psychotherapy," in D. A. Leeming, ed., *Encyclopedia of Psychology and Religion* (Boston: Springer, 2014): 1355–59, doi.org/10.1007/978-1-4614-6086-2_224.

xxi   **Nobel Prize in 1949:** "The Nobel Prize in Physiology or Medicine 1949," Nobelprize.org, www.nobelprize.org/prizes/medicine/1949 /moniz/facts/.

xxi   **we remain in trouble in the emotion department:** U.S. Department of Health and Human Services, "New Surgeon General Advisory Raises

Alarm About the Devastating Impact of the Epidemic of Loneliness and Isolation in the United States," press release, May 3, 2023, www .hhs.gov/about/news/2023/05/03/new-surgeon-general-advisory-raises -alarm-about-devastating-impact-epidemic-loneliness-isolation-united -states.html; "Loneliness Minister: 'It's More Important Than Ever to Take Action,' " gov.uk, June 17, 2021, www.gov.uk/government/news /loneliness-minister-its-more-important-than-ever-to-take-action; "Japan's Parliament Enacts Bill to Tackle Social Isolation," *Japan Times,* May 31, 2023, www.japantimes.co.jp/news/2023/05/31/national /social-isolation-bill/.

xxi   **Bruce Springsteen:** Chloe Melas, "Bruce Springsteen Opens Up About His Battles with Depression: 'I Know I Am Not Completely Well,' " CNN, Nov. 28, 2018, www.cnn.com/2018/11/28/entertainment/bruce -springsteen-mental-health-interview/index.html.

xxi   **A 2020 report:** D. J. Brody and Quiping Gu, "Antidepressant Use Among Adults: United States, 2015–2018," NCHS Data Brief (377) (Sep. 2020): 1–8, www.https://pubmed.ncbi.nlm.nih.gov/33054926/. For an excellent accessible overview of the current state of antidepressant research with references to original research, see Christina Caron, "What You Really Need to Know about Antidepressants," *The New York Times,* April 25, 2024, www.nytimes.com/2024/04/25/well/mind /antidepressants-side-effects-anxiety-stress.html. For another excellent discussion with links to original research, see Dana Smith, "Antidepressants Don't Work the Way Many People Think," *The New York Times,* Nov. 8, 2022, www.nytimes.com/2022/11/08/well/mind /antidepressants-effects-alternatives.html.

xxii   **More than half a billion people:** "Mental Disorders," World Health Organization, June 8, 2022, www.who.int/news-room/fact-sheets /detail/mental-disorders.

xxii   **one trillion dollars:** Lancet Global Health, "Mental Health Matters," *Lancet Global Health* 8, no. 11 (2020), www.thelancet.com /journals/langlo/article/PIIS2214-109X(20)30432-0/fulltext.

xxii   **people who *are* good at managing their emotions:** Terrie Moffitt et al., "A Gradient of Childhood Self-Control Predicts Health, Wealth, and Public Safety," *Proceedings of the National Academy of Sciences* 108, no. 7 (2011): 2693–98, doi:10.1073/pnas.1010076108; Leah S. Richmond-Rakerd et al., "Childhood Self-Control Forecasts the Pace of Midlife Aging and Preparedness for Old Age," *Proceedings of the National Academy of Sciences* 118, no. 3 (2021): e2010211118, doi:10.1073/pnas.2010211118; Benjamin Chapman et al., "High School Personality Traits and 48-Year All-Cause Mortality Risk: Results from a National Sample of 26,845 Baby Boomers," *Journal of Epidemiology* *and Community Health* 73 (2019): 106–10; Markus Jokela et al., "Personality and All-Cause Mortality: Individual-Participant Meta-analysis of 3,947 Deaths in 76,150 Adults," *American Journal of Epidemiology* 178 (2013): 667–75.

xxii   ***Chatter*:** Ethan Kross, *Chatter: The Voice in Our Head, Why It Matters, and How to Harness It* (New York: Crown, 2021).

xxiv   **living in the present:** Several studies show that directing people to think about how they'll feel about something that's bothering them in the future, rather than focusing on how they feel in the moment, alleviates distress. For discussion, see Kross, *Chatter*. For examples of studies that illustrate this point, see Emma Bruehlman-Senecal and Ozlem Ayduk, "This Too Shall Pass: Temporal Distance and the Regulation of Emotional Distress," *Journal of Personality and Social Psychology* 108 (2015): 356–75; and Emma Bruehlman-Senecal, Ozlem Ayduk, and Oliver P. John, "Taking the Long View: Implications of Individual Differences in Temporal Distancing for Affect, Stress Reactivity, and Well-Being," *Journal of Personality and Social Psychology* 111 (2016): 610–35, doi.org/10.1037/pspp0000103.

xxiv   **help us in surprising ways:** For illustrative argument, see Heather C. Lench et al., "Exploring the Toolkit of Emotion: What Do Sadness and Anger Do for Us?," *Social and Personality Psychology Compass* 10, no. 1 (2016): 11–25, doi:10.1111/spc3.12229.

xxv   **no one-size-fits-all solutions:** Bonanno and Burton, "Regulatory Flexibility."

xxv   **instruction manual:** Current therapeutic methods such as cognitive behavioral therapy (CBT) do of course help people regulate their emotions by managing their thoughts, but this book is intended to give everyone, regardless of whether they have access to therapy, the basic building blocks to manage their emotional life. And CBT—as effective as it can be in certain contexts—is ultimately much narrower than this book, focusing as it does on attention and cognition while leaving out the wider universe of sensation, relationships, environments, and culture, all of which we'll explore here, and all of which we can harness for ourselves, outside a therapist's office, anytime we need.

xxvi   **"When Asking 'Why' Does Not Hurt":** Ethan Kross, Ozlem Ayduk, and Walter Mischel, "When Asking Why Does Not Hurt: Distinguishing Rumination from Reflective Processing of Negative Emotions," *Psychological Science* 16 (2005): 709–15.

CHAPTER ONE   *Why We Feel*

3   **Matt Maasdam:** I interviewed Matt over several occasions during the writing of this book to tell his story.

7   **puppet master lurking inside:** The idea that there is some kind of separate entity inside our minds—a miniature version of ourselves buried deep in the brain, sitting at the control panel, pressing buttons and making things happen—popularized in movies like *Inside Out,* is often called the homunculus fallacy and does not reflect how the brain operates.

8   *more than 90 percent of the time*: Debra Trampe et al., "Emotions in Everyday Life," *PLoS ONE* 10, no. 12 (2015): e0145450, doi:10.1371 /journal.pone.0145450. Participants reported experiencing negative emotions 16 percent of the time and positive emotions 41 percent of the time.

8   **like a virus:** Sigal G. Barsade, "The Ripple Effect: Emotional Contagion and Its Influence on Group Behavior," *Administrative Science Quarterly* 47, no. 4 (2002): 644–75, doi:10.2307/3094912; Elaine Hatfield et al., "Emotional Contagion," *Current Directions in Psychological Science* 2, no. 3 (1993): 96–100, doi:10.1111/1467-8721 .ep10770953.

8   **There is a well-circulated quotation:** "They May Forget What You Said, but They Will Never Forget How You Made Them Feel," Quoteinvestigator.com, quoteinvestigator.com/2014/04/06/they-feel/. This quotation is often misattributed to Maya Angelou rather than Carl W. Buehner.

9   **We don't typically:** Debra Trampe et al., "Emotions in Everyday Life."

10   **discrete categories:** Alan S. Cowen and Dacher Keltner, "Self-Report Captures 27 Distinct Categories of Emotion Bridged by Continuous Gradients," *Proceedings of the National Academy of Sciences* 114, no. 38 (2017): E7900–7909, doi:10.1073/pnas.1702247114.

10   **nearly infinite variety:** Lisa Feldman Barrett, *How Emotions Are Made: The Secret Life of the Brain* (Boston: Houghton Mifflin Harcourt, 2017).

10   *Schadenfreude*: Colin Wayne Leach et al., "Malicious Pleasure: Schadenfreude at the Suffering of Another Group," *Journal of Personality and Social Psychology* 84, no. 5 (2003): 932–43, doi:10.1037/0022-3514.84.5.932.

10   **universal responses:** For an overview of the universal stance, see Joseph LeDoux, "Rethinking the Emotional Brain," *Neuron* 73, no. 4 (Feb. 2012): 653–76 doi.org/10.1016/j.neuron.2012.02.004.

10   **utterly unique:** Barrett, *How Emotions Are Made.*

10   **what practically everyone agrees on:** I drew from this excellent synthesis by Klaus Scherer on what different emotions theories agree on: Klaus R. Scherer, "Theory Convergence in Emotion Science Is Timely and Realistic," *Cognition and Emotion* 36, no. 2 (2022): 154–70, doi:10.1080/02699931.2021.1973378.

12   **key building block:** Ibid.; Phoebe C. Ellsworth, "Appraisal Theory: Old and New Questions," *Emotion Review: Journal of the International Society for Research on Emotion 5*, no. 2 (2013): 125–31, doi:10.1177/1754073912463617; Barrett, *How Emotions Are Made.*

14   **positive reframing:** Allison S. Troy et al., "A Person-by-Situation Approach to Emotion Regulation: Cognitive Reappraisal Can Either Help or Hurt, Depending on the Context," *Psychological Science 24*, no. 12 (2013): 2505–14, doi:10.1177/0956797613496434.

14   **essential place in our lives:** Aaron C. Weidman and Ethan Kross, "Examining Emotional Tool Use in Daily Life," *Journal of Personality and Social Psychology 120*, no. 5 (2021): 1344–66, doi:10.1037/pspp 0000292; Heather C. Lench and Zari Koebel Carpenter, "What Do Emotions Do for Us?," in *The Function of Emotions,* ed. Heather C. Lench (New York: Springer, 2018), 1–7; Azim F. Shariff and Jessica L. Tracy, "What Are Emotion Expressions For?," *Current Directions in Psychological Science 20*, no. 6 (2011): 395–99, doi:10.1177/0963721411424739.

15   **anxiety:** Andrew Mathews, "Why Worry? The Cognitive Function of Anxiety," *Behaviour Research and Therapy 28*, no. 6 (1990): 455–68, doi:10.1016/0005-7967(90)90132-3; Jeffrey A. Gray, *The Neuro-psychology of Anxiety: An Enquiry into the Function of the Septo-Hippocampal System* (New York: Oxford University Press, 1982); Lench and Carpenter, "What Do Emotions Do for Us?"

15   **sadness:** For a discussion of the physiological slowing-down effects of sadness, see David Huron, "On the Functions of Sadness and Grief," in Lench, *Function of Emotions,* 59–91. For a discussion of the ways sadness impacts the need to reflect, and its implications for social interactions, see Melissa M. Karnaze and Linda J. Levine, "Sadness, the Architect of Cognitive Change," in Lench, *Function of Emotions,* 45–58.

16   **need support:** S. M. Bell and M. D. Ainsworth, "Infant Crying and Maternal Responsiveness," *Child Development 43*, no. 4 (1972): 1171–90, doi:10.1111/j.1467-8624.1972.tb02075.x; Lawrence Ian Reed and Peter DeScioli, "The Communicative Function of Sad Facial Expressions," *Evolutionary Psychology: An International Journal of Evolutionary Approaches to Psychology and Behavior 15*, no. 1 (2017), doi:10.1177/1474704917700418.

16   **others are more likely to help:** Morteza Dehghani et al., "Interpersonal Effects of Expressed Anger and Sorrow in Morally Charged Negotiation," *Judgment and Decision Making 9*, no. 2 (2014): 104–13, doi:10.1017/s1930297500005477; Ad J. J. M. Vingerhoets and Lauren M. Bylsma, "The Riddle of Human Emotional Crying: A Challenge for Emotion Researchers," *Emotion Review: Journal of the*

International Society for Research on Emotion 8, no. 3 (2016): 207–17, doi:10.1177/1754073915586226.

16   **Envy:** Niels van de Ven et al., "Leveling Up and Down: The Experiences of Benign and Malicious Envy," *Emotion* 9, no. 3 (2009): 419–29, doi:10.1037/a0015669; Jens Lange and Jan Crusius, "The Tango of Two Deadly Sins: The Social-Functional Relation of Envy and Pride," *Journal of Personality and Social Psychology* 109, no. 3 (2015): 453–72, doi:10.1037/pspi0000026; Jens Lange et al., "The Painful Duality of Envy: Evidence for an Integrative Theory and a Meta-analysis on the Relation of Envy and Schadenfreude," *Journal of Personality and Social Psychology* 114, no. 4 (2018): 572–98, doi:10.1037/pspi0000118; Weidman and Kross, "Examining Emotional Tool Use in Daily Life."

16   **Regret:** Daniel H. Pink, *The Power of Regret: How Looking Backward Moves Us Forward* (New York: Random House, 2022).

16   **Guilt:** June Price Tangney and Ronda I. Dearing, *Shame and Guilt* (New York: Guilford Press, 2002); David M. Amodio et al., "A Dynamic Model of Guilt: Implications for Motivation and Self-Regulation in the Context of Prejudice," *Psychological Science* 18, no. 6 (2007): 524–30, doi:10.1111/j.1467-9280.2007.01933.x.

16   **Anger:** Lench and Carpenter, "What Do Emotions Do for Us?"; Heather C. Lench et al., "Exploring the Toolkit of Emotion: What Do Sadness and Anger Do for Us?," *Social and Personality Psychology Compass* 10, no. 1 (2016): 11–25, doi:10.1111/spc3.12229; also see Ira J. Roseman, "Functions of Anger in the Emotion System," in Lench, *Function of Emotions,* 141–73.

16   **Fear:** Parisa Parsafar and Elizabeth L. Davis, "Fear and Anxiety," in Lench, *Function of Emotions,* 9–23; Randolph M. Nesse and Phoebe C. Ellsworth, "Evolution, Emotions, and Emotional Disorders," *American Psychologist* 64, no. 2 (2009): 129–39, doi:10.1037/a0013503; Dean Mobbs et al., "When Fear Is Near: Threat Imminence Elicits Prefrontal-Periaqueductal Gray Shifts in Humans," *Science* 317, no. 5841 (2007): 1079–83, doi:10.1126/science.1144298.

16   **And lust can:** Cindy Hazan and Phillip R. Shaver, "Romantic Love Conceptualized as an Attachment Process," *Journal of Personality and Social Psychology* 52, no. 3 (1987): 511–24, doi.org/10.1037/0022-3514.52.3.511.

16   **In an experiment:** Weidman and Kross, "Examining Emotional Tool Use in Daily Life."

18   **into our very cells:** For a detailed discussion of how the long-term activation of negative emotions can lead to diverse bodily effects, see Kross, *Chatter.*

18   **Dunedin, New Zealand:** Douglas Starr, "Two Psychologists Followed
        1000 New Zealanders for Decades. Here's What They Found About
        How Childhood Shapes Later Life," *Science,* Feb. 1, 2018, www
        .science.org/content/article/two-psychologists-followed-1000-new
        -zealanders-decades-here-s-what-they-found-about-how. For the project
        website, which includes key references, see "The Dunedin Study—
        Dunedin Multidisciplinary Health & Development Research Unit,"
        dunedinstudy.otago.ac.nz/.

20   **a lot about their lives:** Moffitt et al., "Gradient of Childhood Self-
        Control Predicts Health, Wealth, and Public Safety."

20   **Some findings:** Ibid.; Richmond-Rakerd et al., "Childhood Self-
        Control Forecasts the Pace of Midlife Aging and Preparedness for
        Old Age." The links between self-control and brain age and white
        matter hyperintensities became nonsignificant when covariates were
        included in the model, but the rest of the findings remained
        significant.

20   **Some participants got better:** Moffitt et al., "Gradient of Childhood
        Self-Control Predicts Health, Wealth, and Public Safety."

20   **Our ability to regulate our emotions:** Thomas Llewelyn Webb, Eleanor
        Miles, and Paschal Sheeran, "Dealing with Feeling: A Meta-analysis of
        the Effectiveness of Strategies Derived from the Process Model of
        Emotion Regulation," *Psychological Bulletin* 138, no. 4 (2012): 775–
        808, doi.org/10.1037/a0027600; Tal Moran and Tal Eyal, "Emotion
        Regulation by Psychological Distance and Level of Abstraction: Two
        Meta-analyses," *Personality and Social Psychology Review* 26, no. 2
        (2022): 112–59, doi.org/10.1177/10888683211069025; Kateri McRae
        and James J. Gross, "Emotion Regulation," *Emotion* 20, no. 1 (2020):
        1–9, doi.org/10.1037/emo0000703.

21   **Outbursts like these:** Jennifer R. Piazza et al., "Affective Reactivity
        to Daily Stressors and Long-Term Risk of Reporting a Chronic
        Physical Health Condition," *Annals of Behavioral Medicine* 45,
        no. 1 (2013): 110–20, doi:10.1007/s12160-012-9423-0; Susan T. Charles
        et al., "The Wear and Tear of Daily Stressors on Mental Health,"
        Psychological Science 24, no. 5 (2013): 733–41, doi:10.1177/095679761
        2462222.

22   **duration of our emotions:** Philippe Verduyn and Saskia Lavrijsen,
        "Which Emotions Last Longest and Why: The Role of Event
        Importance and Rumination," *Motivation and Emotion* 39, no. 1
        (2015): 119–27, doi:10.1007/s11031-014-9445-y; Philippe Verduyn et al.,
        "Determinants of Emotion Duration and Underlying Psychological
        and Neural Mechanisms," *Emotion Review: Journal of the
        International Society for Research on Emotion* 7, no. 4 (2015): 330–35,
        doi:10.1177/1754073915590618.

22   **a study published in 2015:** Verduyn and Lavrijsen, "Which Emotions Last Longest and Why."

23   **delivering a stressful speech:** Jeremy P. Jamieson et al., "Changing the Conceptualization of Stress in Social Anxiety Disorder: Affective and Physiological Consequences," *Clinical Psychological Science* 1, no. 4 (2013): 363–74, doi:10.1177/2167702613482119; Jeremy P. Jamieson et al., "Improving Acute Stress Responses: The Power of Reappraisal," *Current Directions in Psychological Science* 22, no. 1 (2013): 51–56, doi:10.1177/0963721412461500.

CHAPTER TWO *Can You Really Control Your Emotions?*

29   **When Luisa heard:** I changed the name and minor details to protect the identity of the person described in this story. All other details are true.

32   **In the fall of 2000:** Maya Tamir et al., "Implicit Theories of Emotion: Affective and Social Outcomes Across a Major Life Transition," *Journal of Personality and Social Psychology* 92, no. 4 (2007): 731–44, doi.org/10.1037/0022-3514.92.4.731.

34   **These emotional experiences feel:** A study from 2012 showed that people reported such automatic experiences during half of their waking hours. Wilhelm Hofmann and Lotte Van Dillen, "Desire," *Current Directions in Psychological Science* 21, no. 5 (2012): 317–22, doi.org/10.1177/0963721412453587.

34   **A 2014 study:** Adam S. Radomsky et al., "Part 1—You Can Run but You Can't Hide: Intrusive Thoughts on Six Continents," *Journal of Obsessive-Compulsive and Related Disorders* 3, no. 3 (2014): 269–79, doi:10.1016/j.jocrd.2013.09.002.

34   **What do they look like?:** Christine Purdon and David A. Clark, "Obsessive Intrusive Thoughts in Nonclinical Subjects. Part I. Content and Relation with Depressive, Anxious, and Obsessional Symptoms," *Behaviour Research and Therapy* 31, no. 8 (1993): 713–20, doi.org/10.1016/0005-7967(93)90001-b.

35   **One idea why:** David A. Clark, *Intrusive Thoughts in Clinical Disorders: Theory, Research, and Treatment* (New York, Guilford Press, 2005): 1–29.

37   **Jonathan Cohen:** I'm grateful to a lecture delivered by the Princeton neuroscientist Jonathan Cohen for introducing me to the example of the idea of using the "itch" as an illustration of cognitive control. "Jonathan D Cohen on the Rational Boundedness of Cognitive Control," YouTube, accessed Jan. 9, 2024, www.youtube.com/watch?v=vvwSWkrtQ3s. Also see Xintong Dong and Xinzhong Dong, "Peripheral and Central Mechanisms of Itch," *Neuron* 98, no. 3 (2018): 482–94, pubmed.ncbi.nlm.nih.gov/29723501/.

37   **cognitive control:** Matthew M. Botvinick, et al., "Conflict Monitoring and Cognitive Control," *Psychological Review* 108, no. 3 (2001): 624–52, doi.org/10.1037/0033-295x.108.3.624; E. K. Miller, "The Prefrontal Cortex and Cognitive Control," *Nature Reviews Neuroscience* 1, no. 1 (2000): 59–65, www.ncbi.nlm.nih.gov/pubmed/11252769?dopt =Abstract, doi.org/10.1038/35036228; K. Ochsner and J. J. Gross, "The Cognitive Control of Emotion," *Trends in Cognitive Sciences* 9, no. 5 (May 2005): 242–49, doi.org/10.1016/j.tics.2005.03.010.

38   **Out of all the animals:** Ursula Dicke and Gerhard Roth, "Neuronal Factors Determining High Intelligence," *Philosophical Transactions of the Royal Society B: Biological Sciences* 371, no. 1685 (2016): 20150180, doi.org/10.1098/rstb.2015.0180.

38   **none possesses this capacity:** Ibid.

38   **So why did we evolve:** Francesca De Petrillo et al., "The Evolution of Cognitive Control in Lemurs," *Psychological Science* 33 (2022): 1408–22; Laurie R. Santos and Alexandra G. Rosati, "The Evolutionary Roots of Human Decision-Making," *Annual Review of Psychology* 66 (2015): 321–47; Zhongzheng Fu et al., "Neurophysiological Mechanisms of Error Monitoring in Human and Non-human Primates," *Nature Reviews Neuroscience* 24, no. 3 (2023): 153–72, doi.org/10.1038/s41583-022-00670-w.

39   **ecological intelligence hypothesis:** De Petrillo et al., "Evolution of Cognitive Control in Lemurs"; Richard W. Byrne and Andrew Whiten, *Machiavellian Intelligence: Social Expertise and the Evolution of Intellect in Monkeys, Apes, and Humans* (Oxford: Clarendon Press, 2002); Robin I. Dunbar, "The Social Brain Hypothesis," *Evolutionary Anthropology: Issues, News, and Reviews* 6, no. 5 (1998): 178–90, doi.org/10.1002/(sici)1520-6505(1998)6:5<178::aid-evan5>3.0.co;2-8; Henrike Moll and Michael Tomasello, "Cooperation and Human Cognition: The Vygotskian Intelligence Hypothesis," *Philosophical Transactions of the Royal Society B: Biological Sciences* 362, no. 1480 (2007): 639–48, doi.org/10.1098/rstb.2006.2000; Carel P. Van Schaik and Judith M. Burkart, "Social Learning and Evolution: The Cultural Intelligence Hypothesis," *Philosophical Transactions of the Royal Society B: Biological Sciences* 366, no. 1567 (2011): 1008–16, doi.org /10.1098/rstb.2010.0304; Alexandra G. Rosati, "Foraging Cognition: Reviving the Ecological Intelligence Hypothesis," *Trends in Cognitive Sciences* 21, no. 9 (2017): 691–702, doi.org/10.1016/j.tics.2017.05.011.

40   **One classic study performed in 2004:** Kevin N. Ochsner et al., "For Better or for Worse: Neural Systems Supporting the Cognitive Down- and Up-Regulation of Negative Emotion," *NeuroImage* 23, no. 2 (2004): 483–99. Participants in the study were given two types of instructions to feel better and worse, respectively. I provide examples of

the "situation focused" strategies in the text. They were also provided "self-focused" strategies that involved participants reframing their personal connection to the emotionally evocative images to augment their response.

42   **Albert Bandura recruited participants:** I used several sources to report Bandura's classic research: Albert Bandura, "Applying Theory for Human Betterment," *Perspectives on Psychological Science* 14, no. 1 (2019): 12–15, doi.org/10.1177/1745691618815165; Diane Hamilton, "Moral Disengagement with Dr. Albert Bandura," DrDianeHamilton .com, March 1, 2023, drdianehamilton.com/moral-disengagement-with -dr-albert-bandura/; Angela Duckworth, "Guided Mastery," Character Lab, Oct. 17, 2021, characterlab.org/character-hub/tips/guided -mastery/; Albert Bandura, "Exercise of Control Through Self-Belief," Jan. 4, 1989, garfield.library.upenn.edu/classics1989/A1989U419500001 .pdf; A. Bandura, E. B. Blanchard, and B. Ritter, "Relative efficacy of desensitization and modeling approaches for inducing behavioral, affective, and attitudinal changes," *Journal of Personality and Social Psychology* 13, no. 3 (1969): 173–99.

44   **Decades of research:** Alexander D. Stajkovic and Fred Luthans, "Self-Efficacy and Work-Related Performance: A Meta-analysis," *Psychological Bulletin* 124, no. 2 (1998): 240–61, doi.org /10.1037/0033-2909.124.2.240.

44   **This belief in our own capacities:** Daniel Cervone, "Thinking About Self-Efficacy," *Behavior Modification* 24, no. 1 (2000): 30–56, doi.org /10.1177/0145445500241002.

CHAPTER THREE   *What a 1980s Power Ballad Taught Me*
About Emotion: Sensory Shifters

51   **Boston Consulting Group:** Katharine Shao, "In CNBC Interview, John Legend Talks About His First Post-Penn Job at BCG—and Why He Left It," *Daily Pennsylvanian,* Oct. 12, 2018, www.thedp.com/article /2018/10/john-legend-penn-consulting-bcg-pursue-dreams-music.

52   **temporarily lowering blood pressure:** Martina de Witte et al., "Effects of Music Interventions on Stress-Related Outcomes: A Systematic Review and Two Meta-analyses," *Health Psychology Review* 14, no. 2 (2019): 294–324, doi.org/10.1080/17437199.2019.1627897.

52   **anxiety and paranoia:** E. Stobbe et al., "Birdsongs Alleviate Anxiety and Paranoia in Healthy Participants," *Scientific Reports* 12, no. 1 (2022): 16414, doi.org/10.1038/s41598-022-20841-0.

52   **classic signs of depression:** J. C. Morales-Medina et al., "The Olfactory Bulbectomized Rat as a Model of Depression: The Hippocampal Pathway," *Behavioural Brain Research* 317 (2017): 562–75,

doi.org/10.1016/j.bbr.2016.09.029; Cai Song and Brian E. Leonard, "The Olfactory Bulbectomised Rat as a Model of Depression," *Neuroscience and Biobehavioral Reviews* 29, no. 4–5 (2005): 627–47, doi.org/10.1016/j.neubiorev.2005.03.010.

53 **tool to manage emotions:** There are countless studies linking sensation to emotion change, and certain clinical interventions focus on sensation as well. By emotion regulation frameworks I'm referring to models that provide a scientific blueprint for managing emotions. Despite evidence connecting sensory experiences to emotional phenomenon, leading frameworks are silent on this issue, a point that my colleague Micaela Rodriguez and I made in a paper. See Micaela Rodriguez and Ethan Kross, "Sensory Emotion Regulation," *Trends in Cognitive Sciences* 27, no. 4 (2023): 379–90, doi.org/10.1016/j.tics.2023 .01.0082022.

53 **sang to ourselves and each other:** Jay Schulkin and Greta B. Raglan, "The Evolution of Music and Human Social Capability," *Frontiers in Neuroscience* 8, no. 292 (2014): doi.org/10.3389/fnins.2014.00292.

53 **grieve through music:** G. Casswell, "Beyond Words: Some Uses of Music in the Funeral Setting," *OMEGA - Journal of Death and Dying,* 64(4): 319–34. doi.org/10.2190/OM.64.4.c.

54 **4500 BCE to be exact:** Hazem S. Elshafie and Ippolito Camele, "An Overview of the Biological Effects of Some Mediterranean Essential Oils on Human Health," *BioMed Research International,* Nov. 5, 2017, 1–14, doi.org/10.1155/2017/9268468.

54 **struggling with anxiety:** Ashley J. Farrar and Francisca C. Farrar, "Clinical Aromatherapy," *Nursing Clinics of North America* 55, no. 4 (Dec. 1, 2020): 489–504, www.sciencedirect.com/science/article/pii/S00 29646520300475, doi.org/10.1016/j.cnur.2020.06.015; R. Tisserand, "Essential Oils as Psychotherapeutic Agents," *Springer EBooks* (Jan. 1, 1988):167–81, doi.org/10.1007/978-94-017-2558-3_9. Accessed Aug. 21, 2024.

54 **Egyptian healers:** Donald Bisson, "Reflexology," in *Complementary and Integrative Medicine in Pain Management,* ed. Michael Weintraub et al. (New York: Springer, 2008): 201–14.

54 **sweet, salty, and sour:** Melissa Eisler, "The 6 Tastes of Ayurveda," Chopra, May 16, 2016, chopra.com/articles/the-6-tastes-of-ayurveda.

54 **forty-five thousand years ago:** Avery Hurt, "Why Did Our Paleolithic Ancestors Paint Cave Art?," *Discover Magazine,* Dec. 27, 2022, www .discovermagazine.com/the-sciences/why-did-our-paleolithic-ancestors -paint-cave-art; Sid Perkins, "ScienceShot: Were Most Cave Paintings Done by Women?," *Science,* Oct. 11, 2013, www.science.org/content /article/scienceshot-were-most-cave-paintings-done-women; Adam

Brumm et al., "Oldest Cave Art Found in Sulawesi," *Science Advances* 7, no. 3 (Jan. 1, 2021): eabd4648, advances.sciencemag.org/content/7/3 /eabd4648, doi.org/10.1126/sciadv.abd4648.

54   **unsuspecting customers:** Andrea Cheng, "How a Hotel Gets Its Signature Scent," *Condé Nast Traveler,* Aug. 2, 2019, www.cntraveler .com/story/how-a-hotel-gets-its-signature-scent.

55   **control their mood:** Adam J. Lonsdale and Adrian C. North, "Why Do We Listen to Music? A Uses and Gratifications Analysis," *British Journal of Psychology* 102, no. 1 (2011): 108–34.

55   **my colleague Micaela Rodriguez:** M. Rodriguez and E. Kross, "Harnessing Music as a Tool for Effortless Emotion Regulation" (paper in preparation).

56   **eight out of five hundred students:** Angela L. Duckworth et al., "A Stitch in Time: Strategic Self-Control in High School and College Students," *Journal of Educational Psychology* 108, no. 3 (2016): 329–41, doi.org/10.1037/edu0000062. The numbers quoted in the text were based on unpublished data from this study that Angela Duckworth provided to me for the purpose of this chapter.

56   **T-shirt can reduce cortisol levels:** Marlise K. Hofer et al., "Olfactory Cues from Romantic Partners and Strangers Influence Women's Responses to Stress," *Journal of Personality and Social Psychology* 114, no. 1 (2018): 1–9, doi.org/10.1037/pspa0000110.

56   **petting a dog:** John P. Polheber and Robert L. Matchock, "The Presence of a Dog Attenuates Cortisol and Heart Rate in the Trier Social Stress Test Compared to Human Friends," *Journal of Behavioral Medicine* 37, no. 5 (2013): 860–67, doi.org/10.1007/s10865 -013-9546-1. Also see Emma Ward-Griffin et al., "Petting Away Pre-exam Stress: The Effect of Therapy Dog Sessions on Student Well-Being," *Stress and Health* 34, no. 3 (2018): 468–73, doi.org /10.1002/smi.2804. For broader reviews of the effect of pet therapy (which includes a heavy emphasis on touch) on stress, see Natalie Ein et al., "The Effect of Pet Therapy on the Physiological and Subjective Stress Response: A Meta-analysis," *Stress and Health* 34, no. 4 (2018): 477–89, doi.org/10.1002/smi.2812; and Nancy R. Gee et al., "Dogs Supporting Human Health and Well-Being: A Biopsychosocial Approach," *Frontiers in Veterinary Science* 8 (2021), doi.org/10.3389 /fvets.2021.630465.

56   **hugging a teddy bear:** Sander L. Koole et al., "Embodied Terror Management," *Psychological Science* 25, no. 1 (2013): 30–37, doi .org/10.1177/0956797613483478. For reviews, see Brittany K. Jakubiak and Brooke C. Feeney, "Affectionate Touch to Promote Relational, Psychological, and Physical Well-Being in Adulthood: A Theoretical

Model and Review of the Research," *Personality and Social Psychology Review* 21, no. 3 (2016): 228–52, doi.org/10.1177/1088868316650307; Carissa J. Cascio et al., "Social Touch and Human Development," *Developmental Cognitive Neuroscience* 35 (Feb. 2019): 5–11, doi .org/10.1016/j.dcn.2018.04.009.

56   **dopaminergic pathways in your brain:** Nicole M. Avena, "The Study of Food Addiction Using Animal Models of Binge Eating," *Appetite* 55, no. 3 (2010): 734–37, doi.org/10.1016/j.appet.2010.09.010; P. Rada, N. M. Avena, and B. G. Hoebel, "Daily Bingeing on Sugar Repeatedly Releases Dopamine in the Accumbens Shell," *Neuroscience* 134, no. 3 (2005): 737–44, doi.org/10.1016/j.neuroscience.2005.04.043; Pawel K. Olszewski et al., "Excessive Consumption of Sugar: An Insatiable Drive for Reward," *Current Nutrition Reports* 8, no. 2 (2019): 120–28, doi.org/10.1007/s13668-019-0270-5.

56   **chocolate triggers pleasure:** Laura Fusar-Poli et al., "The Effect of Cocoa-Rich Products on Depression, Anxiety, and Mood: A Systematic Review and Meta-analysis," *Critical Reviews in Food Science and Nutrition* 62, no. 28 (2022): 7905–16, doi.org/10.1080/10408398.2021 .1920570.

56   **pace at which they healed:** Roger S. Ulrich, "View Through a Window May Influence Recovery from Surgery," *Science* 224, no. 4647 (1984): 420–21, doi.org/10.1126/science.6143402. Also see Roger S. Ulrich et al., "Stress Recovery During Exposure to Natural and Urban Environments," *Journal of Environmental Psychology* 11, no. 3 (1991): 201–30, doi.org/10.1016/s0272-4944(05)80184-7; and Daniel K. Brown et al., "Viewing Nature Scenes Positively Affects Recovery of Auton-omic Function Following Acute-Mental Stress," *Environmental Science and Technology* 47, no. 11 (2013): 5562–69, doi.org/10.1021/es305019p.

56   **oldest parts:** Jon H. Kaas, "The Evolution of the Complex Sensory and Motor Systems of the Human Brain," *Brain Research Bulletin* 75, no. 2-4 (Mar. 2008): 384–90, doi.org/10.1016/j.brainresbull.2007.10.009.

57   **Emotions supercharge:** Rodriguez and Kross, "Sensory Emotion Regulation."

58   **due to sensory experience:** For reviews, see Elizabeth A. Kensinger and Jaclyn H. Ford, "Retrieval of Emotional Events from Memory," *Annual Review of Psychology* 71, no. 1 (2019), doi.org/10.1146/annurev-psych -010419-051123; and Linda J. Levine and David A. Pizarro, "Emotion and Memory Research: A Grumpy Overview," *Social Cognition* 22, no. 5 (2004): 530–54, doi.org/10.1521/soco.22.5.530.50767.

58   *In Search of Lost Time:* Marcel Proust, *In Search of Lost Time* (New York: Modern Library, 2003).

59   **increase positive emotion:** As reviewed in Rodriguez and Kross, "Sensory Emotion Regulation"; Jeffrey D. Green et al., "The Proust

Effect: Scents, Food, and Nostalgia," *Current Opinion in Psychology* 50 (April 2023): 101562, doi.org/10.1016/j.copsyc.2023.101562.

59   **trigger fear and stress:** Judith K. Daniels and Eric Vermetten, "Odor-Induced Recall of Emotional Memories in PTSD—Review and New Paradigm for Research," *Experimental Neurology* 284 (Oct. 2016): 168–80, doi.org/10.1016/j.expneurol.2016.08.001.

59   **nausea-inducing radiation:** J. Garcia, D. J. Kimeldorf, and R. A. Koelling, "Conditioned Aversion to Saccharin Resulting from Exposure to Gamma Radiation," *Science* 122, no. 3160 (1955): 157–58, doi.org/10.1126/science.122.3160.157; Carl R. Gustavson et al., "Coyote Predation Control by Aversive Conditioning," *Science* 184, no. 4136 (1974): 581–83, doi.org/10.1126/science.184.4136.581. For review, see John Garcia et al., "A General Theory of Aversion Learning," *Annals of the New York Academy of Sciences* 443, no. 1 (1985): 8–21, doi.org/10.1111/j.1749-6632.1985.tb27060.x.

60   **when they were dumped:** Ethan Kross et al., "Social Rejection Shares Somatosensory Representations with Physical Pain," *Proceedings of the National Academy of Sciences* 108, no. 15 (2011): 6270–75, doi.org/10.1073/pnas.1102693108.

61   **path of least physical and mental effort:** Michael Inzlicht, Amitai Shenhav, and Christopher Y. Olivola, "The Effort Paradox: Effort Is Both Costly and Valued," *Trends in Cognitive Sciences* 22, no. 4 (2018): 337–49.

62   **hip-hop artist Future's "Stick Talk":** Mahita Gajanan, "Here's What Song Michael Phelps Was Listening to When He Made That Face," *Time,* Aug. 29, 2016, time.com/4470449/michael-phelps-olympics-face-future/; Nicole Puglise, "What Is Michael Phelps Listening to on His Trademark Olympics Headphones?," *The Guardian*, Aug. 8, 2016, www.theguardian.com/sport/2016/aug/08/michael-phelps-headphones-music-swimming-olympics-rio#:~:text=.

62   **rub your chin or temples:** N. Ravaja et al., "Feeling Touched: Emotional Modulation of Somatosensory Potentials to Interpersonal Touch," *Scientific Reports* 7, no. 1 (2017): 40504, doi.org/10.1038/srep40504; Jente L. Spille et al., "Cognitive and Emotional Regulation Processes of Spontaneous Facial Self-Touch Are Activated in the First Milliseconds of Touch: Replication of Previous EEG Findings and Further Insights," *Cognitive, Affective, and Behavioral Neuroscience* 22 (2022): 984–1000, doi.org/10.3758/s13415-022-00983-4.

62   **oxytocin and dopamine:** Jakubiak and Feeney, "Affectionate Touch to Promote Relational, Psychological, and Physical Well-Being in Adulthood."

62   **aren't even aware:** I computed the average decoding onset time (Table 2) and rounded to the nearest hundred from this paper: Raphael

Wallroth and Kathrin Ohla, "As Soon as You Taste It: Evidence for Sequential and Parallel Processing of Gustatory Information," *eNeuro* 5, no. 5 (2018): ENEURO.0269-18.2018, doi.org/10.1523/eneuro .0269-18.2018. Also see Rosalind S. E. Carney, "Parallel and Sequential Sequences of Taste Detection and Discrimination in Humans," *eNeuro* 6, no. 1 (2019): ENEURO.0010-19.2019, doi.org/10.1523/ENEURO .0010-19.2019.

63   **viewing scary images:** For review, see Amelia D. Dahlén et al., "Subliminal Emotional Faces Elicit Predominantly Right-Lateralized Amygdala Activation: A Systematic Meta-analysis of fMRI Studies," *Frontiers in Neuroscience* 16 (2022), doi.org/10.3389/fnins.2022 .868366; Arne Öhman et al., "On the Unconscious Subcortical Origin of Human Fear," *Physiology and Behavior* 92, no. 1–2 (2007): 180–85, doi.org/10.1016/j.physbeh.2007.05.057.

63   **In 2022 my team administered:** Chayce Baldwin et al., "Managing Emotions in Everyday Life: Why a Toolbox of Strategies Matters" (University of Michigan document).

63   **Using effortful emotion regulation tools:** Ariana Orvell et al., "Does Distanced Self-Talk Facilitate Emotion Regulation Across a Range of Emotionally Intense Experiences?," *Clinical Psychological Science* 9, no. 1 (2021): 68–78.

63   **even if you're distracted:** As discussed in Rodriguez and Kross. For examples, see Iris Duif et al., "Effects of Distraction on Taste-Related Neural Processing: A Cross-Sectional fMRI Study," *American Journal of Clinical Nutrition* 111, no. 5 (2020), doi.org/10.1093/ajcn/nqaa032.

63   **can be soothing:** Spille et al., "Cognitive and Emotional Regulation Processes of Spontaneous Facial Self-Touch Are Activated in the First Milliseconds of Touch."

64   **emotional congruency effect:** Patrick G. Hunter et al., "Misery Loves Company: Mood-Congruent Emotional Responding to Music," *Emotion* 11, no. 5 (2011): 1068–72, doi.org/10.1037/a0023749.

65   **risky sexual behavior:** Tatjana van Strien and Machteld A. Ouwens, "Effects of Distress, Alexithymia, and Impulsivity on Eating," *Eating Behaviors* 8, no. 2 (2007): 251–57, doi.org/10.1016/j.eatbeh.2006 .06.004; Catherine Potard, Robert Courtois, and Emmanuel Rusch, "The Influence of Peers on Risky Sexual Behaviour During Adolescence," *European Journal of Contraception and Reproductive Health Care* 13, no. 3 (2008): 264–70, doi.org/10.1080/13625180 802273530.

66   **first sense to be developed:** Mark Paterson, *The Senses of Touch: Haptics, Affects, and Technologies* (London: Bloomsbury Academic, 2013); Tiffany Field, *Touch* (Cambridge, Mass.: MIT Press, 2014).

67   **"loneliness epidemic":** U.S. Department of Health and Human Services, "New Surgeon General Advisory Raises Alarm about the Devastating Impact of the Epidemic of Loneliness and Isolation in the United States," May 3, 2023, www.hhs.gov/about/news/2023/05/03/new -surgeon-general-advisory-raises-alarm-about-devastating-impact -epidemic-loneliness-isolation-united-states.html.

67   **In a pilot study we ran:** These testimonials were collected during an exploratory study in which we asked participants to describe how they coped with feeling lonely.

CHAPTER FOUR   *The Myth of Universal Approach: Attention Shifters*

71   **In a video from the mid 1990s:** Dimitro, interview by Goldman, collections.ushmm.org/search/catalog/irn504865.

73   **forget just how much it matters:** For reviews, see M. I. Posner, "Attention: The Mechanisms of Consciousness," *Proceedings of the National Academy of Sciences* 91, no. 16 (1994): 7398–403, doi.org /10.1073/pnas.91.16.7398; Freek van Ede and Anna C. Nobre, "Turning Attention Inside Out: How Working Memory Serves Behavior," *Annual Review of Psychology* 74, no. 1 (2022), doi.org/10.1146 /annurev-psych-021422-041757.

75   **I came across a classic paper:** E. B. Foa and M. J. Kozak, "Emotional Processing of Fear: Exposure to Corrective Information," *Psychological Bulletin* 99, no. 1 (1986): 20–35, pubmed.ncbi.nlm.nih.gov/2871574/.

76   **trying to make meaning:** Ethan Kross, *Chatter.*

78   **"Dennis was bizarre":** Drew Shiller, "Steve Kerr Explains Why Phil Jackson Let Dennis Rodman Go to Las Vegas," NBC Sports, April 27, 2020, www.nbcsportsbayarea.com/nba/golden-state-warriors/steve -kerr-explains-why-phil-jackson-let-dennis-rodman-go-to-las-vegas /1364859/.

79   **"I think Phil":** Jason Heir, *The Last Dance*, episode 3 (Netflix: April 2020).

79   **fourth game:** Justin Barrasso, "Rodman Once Skipped Practice During Finals for WCW Gig," *Sports Illustrated*, April 27, 2020, www.si.com/ wrestling/2020/04/27/dennis-rodman-wcw-nitro -1998-nba-finals-practice.

81   **misperception that when it comes to emotions, avoidance is always toxic persists:** Stefan G. Hofmann and Aleena C. Hay, "Rethinking Avoidance: Toward a Balanced Approach to Avoidance in Treating Anxiety Disorders," *Journal of Anxiety Disorders* 55 (2018): 14–21.

81   **perpetuating our distress:** G. A. Bonanno et al., "When Avoiding Unpleasant Emotions Might Not Be Such a Bad Thing: Verbal-

Autonomic Response Dissociation and Midlife Conjugal Bereavement," *Journal of Personality and Social Psychology* 69, no. 5 (1995).

81 **"avoidance coping"**: For a review, see Charles J. Holahan et al., "Stress Generation, Avoidance Coping, and Depressive Symptoms: A 10-Year Model," *Journal of Consulting and Clinical Psychology* 73, no. 4 (2005): 658–66, doi.org/10.1037/0022-006x.73.4.658.

82 **psychological immune system:** Kross, *Chatter;* D. T. Gilbert et al., "Immune Neglect: A Source of Durability Bias in Affective Forecasting," *Journal of Personality and Social Psychology* 75, no. 3 (1998): 617–38, www.ncbi.nlm.nih.gov/pubmed/9781405, doi.org /10.1037//0022-3514.75.3.617.

83 **"time heals all wounds":** Michael J. A. Wohl and April L. McGrath, "The Perception of Time Heals All Wounds: Temporal Distance Affects Willingness to Forgive Following an Interpersonal Transgression," *Personality and Social Psychology Bulletin* 33.7 (2007): 1023–35; Emma Bruehlman-Senecal, Özlem Ayduk, and Oliver P. John, "Taking the Long View: Implications of Individual Differences in Temporal Distancing for Affect, Stress Reactivity, and Well-Being," *Journal of Personality and Social Psychology* 111.4 (2016): 610–35.

83 **chronic approach:** An example of what you might consider chronic approach is rumination, where you focus over and over on the problems you're experiencing without making progress working-through them; see Ethan Kross, *Chatter.*

83 **Studies show that *on average*:** Amelia Aldao et al., "Emotion-Regulation Strategies across Psychopathology: A Meta-Analytic Review," *Clinical Psychology Review* 30, no. 2 (March 2010): 217–37, pubmed .ncbi.nlm.nih.gov/20015584/, doi.org/10.1016/j.cpr.2009.11.004.

85 **best able to cope:** George A. Bonanno et al., "The Importance of Being Flexible," *Psychological Science* 15.7 (2004): 482–87.

86 **better psychological health:** Cecilia Cheng, Hi-Po Bobo Lau, and Man-Pui Sally Chan, "Coping Flexibility and Psychological Adjustment to Stressful Life Changes: A Meta-analytic Review," *Psychological Bulletin* 140, no. 6 (2014): 1582–607.

CHAPTER FIVE   *"Easier F***ing Said Than Done": Perspective Shifters*

93 **The American astronaut Jerry Linenger:** To tell Jerry's story, I relied on an interview I performed on February 23, 2013, and his memoir, Jerry M. Linenger, *Off the Planet: Surviving Five Perilous Months Aboard the Space Station Mir* (New York: McGraw-Hill, 2000).

94 **All the breathable air:** Anna Gosline, "Survival in Space Unprotected Is Possible—Briefly," *Scientific American,* Feb. 14, 2008.

97   **cognitive behavioral therapy:** Changing your beliefs to regulate your emotions has been around for a while. In both Western and Eastern ancient philosophy there is a long tradition espousing the benefits of changing your mind to change how you feel. The Greek Stoic Epictetus once famously said, "Men are disturbed not by things but by the views which they take of them." It took almost two millennia, but eventually science caught up with Epictetus and his buddies. Psychologists in the nineteenth century were initially taken with the idea that unconscious forces give rise to our emotions—calling Dr. Freud, here. Then in the next generation there was a rejection of that and a focus on what was observable—think Pavlov's dogs. Finally, came the "cognitive revolution" that swept through the world of psychology in the 1970s and has maintained its primacy ever since.

99   **trap of reframing *negatively*:** Some people think that the key to reframing is to positively reinterpret your feelings. While there are many benefits associated with such reframes, positive reframing is only one example of a helpful cognitive shift. You can also reframe your circumstances by adopting a more sober detached perspective that allows you to take an objective view. In this instance you're still thinking about a negative situation—you're not making lemonade out of lemons—but you're able to wade through the negativity in a more productive way that ultimately helps you out.

100   **peer into the brains of worriers:** Debra A. Bangasser and Amelia Cuarenta, "Sex Differences in Anxiety and Depression: Circuits and Mechanisms," *Nature Reviews Neuroscience,* Sept. 20, 2021, doi.org /10.1038/s41583-021-00513-0; Liana S. Leach et al., "Gender Differences in Depression and Anxiety Across the Adult Lifespan: The Role of Psychosocial Mediators," *Social Psychiatry and Psychiatric Epidemiology* 43, no. 12 (2008): 983–98, doi.org/10.1007/s00127-008 -0388-z; Carmen P. McLean and Emily R. Anderson, "Brave Men and Timid Women? A Review of the Gender Differences in Fear and Anxiety," *Clinical Psychology Review* 29, no. 6 (2009): 496–505, doi .org/10.1016/j.cpr.2009.05.003.

101   **brain readouts:** In technical terms they showed heightened levels of activity on the stimulus preceding negativity, a neurophysiological waveform that tracks effortful cognitive processing, and considerably more activity on the late positive potential, a neural waveform that tracks self-referential emotional reactivity. Jason S. Moser et al., "Neural Markers of Positive Reappraisal and Their Associations with Trait Reappraisal and Worry," *Journal of Abnormal Psychology* 123, no. 1 (2014): 91–105, doi.org/10.1037/a0035817.

102   **In one study, researchers analyzed:** Sarah Seraj et al., "Language Left Behind on Social Media Exposes the Emotional and Cognitive Costs of

a Romantic Breakup," *Proceedings of the National Academy of Sciences* 118, no. 7 (2021), doi.org/10.1073/pnas.2017154118.

102 **Another, complementary study:** Johannes C. Eichstaedt et al., "Facebook Language Predicts Depression in Medical Records," *Proceedings of the National Academy of Sciences* 115, no. 44 (2018): 11203–8, doi.org/10.1073/pnas.1802331115.

103 **"dramatic loss of prefrontal cognitive abilities":** Amy F. T. Arnsten, "Stress Signalling Pathways That Impair Prefrontal Cortex Structure and Function," *Nature Reviews Neuroscience* 10, no. 6 (June 2009): 410–22, www.ncbi.nlm.nih.gov/pmc/articles/PMC2907136/#:~:text =The%20prefrontal%20cortex%20(PFC)%20intelligently,brain%20 regions%20(BOX%201), doi.org/10.1038/nrn2648.

104 **Research shows that a link:** Jean-Marc Dewaele, "The Emotional Force of Swearwords and Taboo Words in the Speech of Multilinguals," *Journal of Multilingual and Multicultural Development* 25, no. 2–3 (2004): 204–22, doi.org/10.1080/01434630408666529; Jean-Marc Dewaele, "The Emotional Weight of *I Love You* in Multilinguals' Languages," *Journal of Pragmatics* 40, no. 10 (2008): 1753–80, doi.org /10.1016/j.pragma.2008.03.002; Fernando Gonzalez-Reigosa, "The Anxiety-Arousing Effect of Taboo Words in Bilinguals," in *Cross-Cultural Anxiety,* ed. Charles D. Spielberger and Rogelio Diaz-Guerrero (Washington, D.C.: Hemisphere, 1976), 89–105; Catherine L. Caldwell-Harris et al., "Physiological Reactivity to Emotional Phrases in Mandarin–English Bilinguals," *International Journal of Bilingualism* 15, no. 3 (2011): 329–52, doi.org/10.1177/1367006910379 262; Jennifer Suzanne Schwanberg, "Does Language of Retrieval Affect the Remembering of Trauma?," *Journal of Trauma and Dissociation* 11, no. 1 (2010): 44–56, doi:10.1080/15299730903143550; Catherine L. Harris, "Bilingual Speakers in the Lab: Psychophysiological Measures of Emotional Reactivity," *Journal of Multilingual and Multicultural Development* 25, no. 2–3 (2004): 223–47, doi.org/10.1080/01434630408 666530; Catherine L. Harris, Ayşe Ayçiçeği, and Jean Berko Gleason, "Taboo Words and Reprimands Elicit Greater Autonomic Reactivity in a First Language Than in a Second Language," *Applied Psycholinguistics* 24, no. 4 (2003): 561–79, doi.org/10.1017/S0142716403000286; Sayuri Hayakawa et al., "Thinking More or Feeling Less? Explaining the Foreign-Language Effect on Moral Judgment," *Psychological Science* 28, no. 10 (Aug. 14, 2017): 1387–97, doi.org/10.1177/0956797617720944; Boaz Keysar et al., "The Foreign-Language Effect: Thinking in a Foreign Tongue Reduces Decision Biases on Behalf Of: Association for Psychological Science," *Sage Journals* 23, no. 6 (2012), doi .org/10.1177/0956797611432178.

105 **Novak Djokovic:** www.facebook.com/TennisMajors; "Mamba

Mentality: The Exact Words Djokovic Told Himself in the Mirror," *Tennis Majors,* July 6, 2022, www.tennismajors.com/wimbledon-news /mamba-mentality-the-exact-words-djokovic-told-himself-in-the -mirror-610966.html.

106 **Solomon's paradox:** Igor Grossmann and Ethan Kross, "Exploring Solomon's Paradox: Self-Distancing Eliminates the Self-Other Asymmetry in Wise Reasoning About Close Relationships in Younger and Older Adults," *Psychological Science* 25, no. 8 (2014): 1571–80, doi.org/10.1177/0956797614535400.

106 **Using the word "you":** For an overview, see Ariana Orvell et al., "What 'You' and 'We' Say About Me: How Small Shifts in Language Reveal and Empower Fundamental Shifts in Perspective," *Social and Personality Psychology Compass,* April 6, 2022, doi.org/10.1111/spc3 .12665; Ariana Orvell et al., "Linguistic Shifts: A Relatively Effortless Route to Emotion Regulation?," *Current Directions in Psychological Science* 28, no. 6 (2019): 567–73, doi.org/10.1177/0963721419861411. For an overview of research on self-distancing and emotion regulation effects, see Moran and Eyal, "Emotion Regulation by Psychological Distance and Level of Abstraction"; Ethan Kross et al., "Self-Reflection at Work: Why It Matters and How to Harness Its Potential and Avoid Its Pitfalls," *Annual Review of Organizational Psychology and Organizational Behavior* 10, no. 1 (2023): 441–64, doi.org/10.1146 /annurev-orgpsych-031921-024406.

107 **same way that "you" does:** Research hasn't systematically explored the role that "they" plays as a linguistic shifter, but from a theoretical perspective we'd expect it to have similar benefits to using "you" or "he" or "she."

108 **In one pair of neuroscience experiments:** Christopher T. Webster et al., "An Event-Related Potential Investigation of Distanced Self-Talk: Replication and Comparison to Detached Reappraisal," *International Journal of Psychophysiology* 177 (July 2022); Jason S. Moser, "Third-Person Self-Talk Facilitates Emotion Regulation Without Engaging Cognitive Control: Converging Evidence from ERP and fMRI," *Scientific Reports* 7, no. 1 (2017), doi.org/10.1038/s41598-017-04047-3.

108 **Other research:** Erik C. Nook et al., "A Linguistic Signature of Psychological Distancing in Emotion Regulation," *Journal of Experimental Psychology: General* 146, no. 3 (2017): 337–46, doi .org/10.1037/xge0000263; Erik C. Nook et al., "Use of Linguistic Distancing and Cognitive Reappraisal Strategies During Emotion Regulation in Children, Adolescents, and Young Adults," *Emotion* 20, no. 4 (2020): 525–40, doi.org/10.1037/emo0000570.

108 **Across these contexts:** Orvell et al., "Does Distanced Self-Talk Facilitate Emotion Regulation Across a Range of Emotionally Intense

Experiences?"; Ethan Kross et al., "Self-Talk as a Regulatory Mechanism: How You Do It Matters," *Journal of Personality and Social Psychology* 106, no. 2 (2014): 304–24, doi.org/10.1037/a0035173; Sanda Dolcos and Dolores Albarracin, "The Inner Speech of Behavioral Regulation: Intentions and Task Performance Strengthen When You Talk to Yourself as a You," *European Journal of Social Psychology* 44, no. 6 (2014): 636–42, doi.org/10.1002/ejsp.2048; Ethan Zell et al., "Splitting of the Mind," *Social Psychological and Personality Science* 3, no. 5 (2011): 549–55, doi.org/10.1177/1948550611430164.

109   **One study explored:** Erik C. Nook et al., "Linguistic Measures of Psychological Distance Track Symptom Levels and Treatment Outcomes in a Large Set of Psychotherapy Transcripts," *Proceedings of the National Academy of Sciences* 119, no. 13 (2022), doi.org/10.1073/pnas.2114737119.

109   **improved how wisely:** Igor Grossmann et al., "Training for Wisdom: The Distanced-Self-Reflection Diary Method," *Psychological Science* 32, no. 3 (2021): 381–94, doi.org/10.1177/0956797620969170.

112   **Using mental time travel:** Bruehlman-Senecal and Ayduk, "This Too Shall Pass"; Bruehlman-Senecal, Ayduk, and John, "Taking the Long View."

113   **Vic Strecher:** Victor J. Strecher, *Life on Purpose: How Living for What Matters Most Changes Everything* (New York: HarperOne, 2016).

114   **Paul Kalanithi:** Paul Kalanithi, *When Breath Becomes Air* (New York: Random House, 2016).

115   **Viktor Frankl:** Viktor E. Frankl, *Man's Search for Meaning* (Boston: Beacon Press, 1946).

CHAPTER SIX   *Hidden in Plain Sight: Space Shifters*

119   **Laurie Santos:** I interviewed Laurie on two occasions to tell her story.

119   **The problem was:** This is a great illustration of Solomon's paradox, our tendency to be able to give others advice better than we can give ourselves.

123   **regional differences across China:** T. Talhelm et al., "Large-Scale Psychological Differences Within China Explained by Rice Versus Wheat Agriculture," *Science* 344, no. 6184 (2014): 603–8, doi.org/10.1126/science.1246850.

124   **envy-inducing social comparisons:** Cheol-Sung Lee et al., "People in Historically Rice-Farming Areas Are Less Happy and Socially Compare More Than People in Wheat-Farming Areas," *Journal of Personality and Social Psychology* 124, no. 5 (2023): 935–57.

125   **When Sean:** I changed the name and minor details to protect the identity of the person described in this story. All other details are true.

128   **sense of awe:** Kross, *Chatter;* Gregory N. Bratman et al., "Nature and Mental Health: An Ecosystem Service Perspective," *Science Advances* 5, no. 7 (2019), doi.org/10.1126/sciadv.aax0903; Dacher Keltner, *Awe* (New York: Penguin Press, 2023).

128   **place attachment:** Robert Gifford, "Environmental Psychology Matters," *Annual Review of Psychology* 65, no. 1 (2014): 541–79, doi.org/10.1146/annurev-psych-010213-115048; Leila Scannell and Robert Gifford, "Defining Place Attachment: A Tripartite Organizing Framework," *Journal of Environmental Psychology* 30, no. 1 (2010): 1–10, doi.org/10.1016/j.jenvp.2009.09.006; Maria Lewicka, "Place Attachment: How Far Have We Come in the Last 40 Years?," *Journal of Environmental Psychology* 31, no. 3 (2011): 207–30, doi.org/10.1016/j.jenvp.2010.10.001.

128   **Call it a happy place:** Kathleen Wolf, "Place Attachment and Meaning," Green Cities: Good Health, College of the Environment, University of Washington, 2014, depts.washington.edu/hhwb/Thm_Place.html.

128   **attachments to our caregivers:** Hazan and Shaver, "Romantic Love Conceptualized as an Attachment Process"; Mario Mikulincer and Phillip R. Shaver, *Attachment in Adulthood: Structure, Dynamics, and Change,* 2nd ed. (New York: Guilford Press, 2016).

129   **climate justice advocates:** Jerry J. Vaske and Katherine C. Kobrin, "Place Attachment and Environmentally Responsible Behavior," *Journal of Environmental Education* 32, no. 4 (2001): 16–21.

133   **situation modification:** James J. Gross, "The Emerging Field of Emotion Regulation: An Integrative Review," *Review of General Psychology* 2, no. 3 (1998): 271–99, doi.org/10.1037//1089-2680.2.3.271. Also see Angela L. Duckworth et al., "Situational Strategies for Self-Control," *Perspectives on Psychological Science* 11, no. 1 (2016): 35–55, doi.org/10.1177/1745691615623247; Walter Mischel et al., "Cognitive and Attentional Mechanisms in Delay of Gratification," *Journal of Personality and Social Psychology* 21, no. 2 (1972): 204–18, doi.org/10.1037/h0032198; Walter Mischel, *The Marshmallow Test: Understanding Self-Control and How to Master It* (London: Corgi Books, 2015).

134   **how high school students control:** Duckworth et al., "Stitch in Time."

136   **view of self-control:** Kentaro Fujita et al., "Smarter, Not Harder: A Toolbox Approach to Enhancing Self-Control," *Policy Insights from the Behavioral and Brain Sciences* 7, no. 2 (2020): 149–56, doi.org/10.1177/2372732220941242; Denise T. D. de Ridder et al., "Taking Stock of Self-Control: A Meta-analysis of How Trait Self-Control Relates to

a Wide Range of Behaviors," *Personality and Social Psychology Review* 16, no. 1 (2012): 76–99, doi.org/10.1177/1088868311418749.

136   **help people manage emotional pain:** Emre Selcuk et al., "Mental Representations of Attachment Figures Facilitate Recovery Following Upsetting Autobiographical Memory Recall," *Journal of Personality and Social Psychology* 103, no. 2 (2012): 362–78, doi.org/10.1037/a0028125.

139   **best predictors of happiness:** Ed Diener and Martin E. P. Seligman, "Very Happy People," *Psychological Science* 13, no. 1 (2002): 81–84, doi.org/10.1111/1467-9280.00415; Robert Waldinger and Marc Schulz, *The Good Life* (New York: Simon & Schuster, 2023).

139   **propinquity effect:** Leon Festinger, Stanley Schachter, and Kurt Back, *Social Pressures in Informal Groups: A Study of Human Factors in Housing* (New York: Harper, 1950).

CHAPTER SEVEN   *Catching a Feeling: Relationship Shifters*

142   **Their desire to belong:** Roy F. Baumeister and Mark R. Leary, "The Need to Belong: Desire for Interpersonal Attachments as a Fundamental Human Motivation," *Psychological Bulletin* 117, no. 3 (1995): 497–529, doi.org/10.1037/0033-2909.117.3.497.

143   **The rapid transmission of emotion:** Sigal Barsade, "The Contagion We Can Control," *Harvard Business Review,* March 26, 2020, hbr.org/2020 /03/the-contagion-we-can-control; Barsade, "Ripple Effect."

143   **just as potent:** Sigal Barsade et al., "Emotional Contagion in Organizational Life," *Research in Organizational Behavior* 38 (2018): 137–51, doi.org/10.1016/j.riob.2018.11.005.

144   **"catch a feeling":** Alison L. Hill et al., "Emotions as Infectious Diseases in a Large Social Network: The SISa Model," *Proceedings of the Royal Society B: Biological Sciences* 277, no. 1701 (2010): 3827–35, doi.org/10.1098/rspb.2010.1217.

144   **since time immemorial:** For classic references, see Gustave Le Bon, *The Crowd: A Study of the Popular Mind* (New York: Viking Press, 1960); and Elaine Hatfield, John T. Cacioppo, and Richard L. Rapson, *Emotional Contagion* (Cambridge, U.K.: Cambridge University Press, 2003). For a contemporary overview, see Barsade et al., "Emotional Contagion in Organizational Life."

144   **you've seen this firsthand:** Barsade, "Contagion We Can Control"; Abraham Sagi and Martin L. Hoffman, "Empathic Distress in the Newborn," *Developmental Psychology* 12, no. 2 (1976): 175–76, doi.org/10.1037/0012-1649.12.2.175. For a review, see Korrina A. Duffy and Tanya L. Chartrand, "Mimicry: Causes and Consequences," *Current Opinion in Behavioral Sciences* 3 (June 2015): 112–16, doi.org/10.1016/j.cobeha.2015.03.002; Susan S. Jones, "Imitation in

Infancy," *Psychological Science* 18, no. 7 (2007): 593–99, doi.org
/10.1111/j.1467-9280.2007.01945.x.

144 **mimicry within seconds:** For an excellent discussion of the boundary
conditions surrounding emotional contagion, see Guillaume Dezecache
et al., "Emotional Contagion: Its Scope and Limits," *Trends in
Cognitive Sciences* 19, no. 6 (2015): 297–99, doi.org/10.1016/j.tics.2015
.03.011. Also see Patrick Bourgeois and Ursula Hess, "The Impact of
Social Context on Mimicry," *Biological Psychology* 7, no. 3 (2008): 343–52.

144 **When we mirror others:** Scientists have documented conditions that
facilitate versus inhibit mimicry. For a concise discussion, see Duffy
and Chartrand, "Mimicry."

144 **completely unaware of it:** For a discussion of the conscious and
unconscious pathways to emotional contagion, see Barsade et al.,
"Emotional Contagion in Organizational Life."

144 **cascade within you:** Gerben A. van Kleef and Stéphane Côté, "The
Social Effects of Emotions," *Annual Review of Psychology* 73 (Jan.
2022): 629–58.

144 **susceptibility to burnout:** Arnold B. Bakker et al., "Burnout Contagion
Among General Practitioners," *Journal of Social and Clinical
Psychology* 20, no. 1 (2001): 82–98. doi.org/10.1521/jscp.20.1.82.22251;
Willem Verbeke, "Individual Differences in Emotional Contagion of
Salespersons: Its Effect on Performance and Burnout," *Psychology &
Marketing* 14, no. 6 (1997): 617–36, doi.org/10.1002/(sici)1520-6793
(199709)14:6<617::aid-mar6>3.0.co;2-a.

144 **how well negotiations go:** Barsade et al., "Emotional Contagion in
Organizational Life."

147 **unproductive co-rumination:** Amanda J. Rose, "The Costs and Benefits
of Co-rumination," *Child Development Perspectives* 15, no. 3 (2021):
176–81, doi.org/10.1111/cdep.12419; David S. Lee et al., "When
Chatting About Negative Experiences Helps—and When It Hurts:
Distinguishing Adaptive Versus Maladaptive Social Support in
Computer-Mediated Communication," *Emotion* 20, no. 3 (2019),
doi.org/10.1037/emo0000555.

147 **rehashing their feelings:** Bernard Rimé et al., "Intrapersonal,
Interpersonal, and Social Outcomes of the Social Sharing of Emotion,"
Current Opinion in Psychology 31 (Feb. 2020): 127–34, doi.org/10.1016
/j.copsyc.2019.08.024; Lisanne S. Pauw et al., "I Hear You (Not):
Sharers' Expressions and Listeners' Inferences of the Need for Support
in Response to Negative Emotions," *Cognition and Emotion* 33, no. 6
(2018): 1129–43, doi.org/10.1080/02699931.2018.1536036; Bernard
Rimé, "Emotion Elicits the Social Sharing of Emotion: Theory and
Empirical Review," *Emotion Review* 1, no. 1 (Jan. 2009): 60–85,
doi.org/10.1177/1754073908097189.

147   **early attachment instincts:** Mario Mikulineer and Phillip R. Shaver, "An Attachment and Behavioral Systems Perspective on Social Support," *Journal of Social and Personal Relationships* 26, no. 1 (2009): 7–19, doi.org/10.1177/0265407509105518.

148   **literally be paid:** Diana I. Tamir and Jason P. Mitchell, "Disclosing Information About the Self Is Intrinsically Rewarding," *Proceedings of the National Academy of Sciences* 109, no. 21 (2012): 8038–43.

149   **the more intense the emotion:** Bernard Rimé, "Emotion Elicits the Social Sharing of Emotion: Theory and Empirical Review," *Emotion Review* 1, no. 1 (Jan. 2009): 60–85, doi.org/10.1177/1754073908097189.

149   **from a broader perspective:** Kross, *Chatter;* Lee et al., "When Chatting About Negative Experiences Helps—and When It Hurts"; Bernard Rimé, "Emotion Elicits the Social Sharing of Emotion: Theory and Empirical Review," *Emotion Review* 1, no. 1 (2009): 60–85, doi.org/10.1177/1754073908097189.

154   **positive mood declined over time:** Ethan Kross et al., "Facebook Use Predicts Declines in Subjective Well-Being in Young Adults," *PLoS ONE* 8, no. 8 (2013), doi.org/10.1371/journal.pone.0069841.

154   **predicted declines in their well-being:** Philippe Verduyn et al., "Passive Facebook Usage Undermines Affective Well-Being: Experimental and Longitudinal Evidence," *Journal of Experimental Psychology: General* 144, no. 2 (2015): 480–88, pubmed.ncbi.nlm.nih.gov/25706656/.

154   **three billion people:** "Most Popular Social Networks Worldwide as of April 2024, by Number of Monthly Active Users," Statista, 2024, www.statista.com/statistics/272014/global-social-networks-ranked-by-number-of-users/.

154   **digital social comparison:** Moira Burke et al., "Social Comparison and Facebook: Feedback, Positivity, and Opportunities for Comparison," *Proceedings of the 2020 CHI Conference on Human Factors in Computing Systems* (2020): 1–13, reviewed in Ethan Kross et al., "Social Media and Well-Being: Pitfalls, Progress, and Next Steps," *Trends in Cognitive Sciences* 25, no. 1 (2020), doi.org/10.1016/j.tics.2020.10.005.

154   **unhelpful comparisons:** "Daily Time Spent on Social Networking by Internet Users Worldwide from 2012 to 2022," Statista, March 21, 2022, www.statista.com/statistics/433871/daily-social-media-usage-worldwide/.

154   **negative experiences loom larger:** Roy F. Baumeister et al., "Bad Is Stronger Than Good," *Review of General Psychology* 5, no. 4 (2001): 323–70. Also see Daniel Kahneman and Amos Tversky, "Prospect

Theory: An Analysis of Decision Under Risk," *Econometrica* 47, no. 2 (1979): 263–92, www.jstor.org/stable/1914185.

154   **people who should know better:** Jerry Suls et al., "Social Comparison: Why, with Whom, and with What Effect?," *Current Directions in Psychological Science* 11, no. 5 (2002): 159–63, doi.org/10.1111/1467 -8721.00191.

154   **baked into our brains:** K. Fliessbach et al., "Social Comparison Affects Reward-Related Brain Activity in the Human Ventral Striatum," *Science* 318, no. 5854 (2007): 1305–8, doi.org/10.1126/science.1145876.

155   **regardless of our income bracket:** For a review, see Verduyn et al., "Social Comparison on Social Networking Sites"; Peter R. Blake and Katherine McAuliffe, " 'I Had So Much It Didn't Seem Fair': Eight- Year-Olds Reject Two Forms of Inequity," *Cognition* 120, no. 2 (2011): 215–24, doi.org/10.1016/j.cognition.2011.04.006; Vanessa LoBue et al., "When Getting Something Good Is Bad: Even Three-Year-Olds React to Inequality," *Social Development* 20, no. 1 (2010): 154–70, doi.org /10.1111/j.1467-9507.2009.00560.x; Ernst Fehr et al., "Egalitarianism in Young Children," *Nature* 454, no. 7208 (2008): 1079–83, doi.org /10.1038/nature07155; Joyce F. Benenson et al., "Do Young Children Understand Relative Value Comparisons?," *PLoS ONE* 10, no. 4 (2015): e0122215, doi.org/10.1371/journal.pone.0122215.

155   **weighing how we are doing:** Matthew Baldwin and Thomas Mussweiler, "The Culture of Social Comparison," *Proceedings of the National Academy of Sciences* 115, no. 39 (2018): E9067–74, doi.org /10.1073/pnas.1721555115; Thomas Mussweiler, "Comparison Processes in Social Judgment: Mechanisms and Consequences," *Psychological Review* 110, no. 3 (2003): 472–89, doi.org/10.1037/0033 -295x.110.3.472; Suls et al., "Social Comparison"; Abraham Tesser, "Toward a Self-Evaluation Maintenance Model of Social Behavior," in *Advances in Experimental Social Psychology,* ed. Leonard Berkowitz (New York: Academic, 1988), 21:181–227; Paul Gilbert et al., "Social Comparison, Social Attractiveness, and Evolution: How Might They Be Related?," *New Ideas in Psychology* 13, no. 2 (1995): 149–65, doi .org/10.1016/0732-118x(95)00002-x.

155   **college nemeses:** For a discussion of spontaneous versus effortful social comparisons, see Baldwin and Mussweiler, "Culture of Social Comparison."

155   **direct our behavior:** Leon Festinger, "A Theory of Social Comparison Processes," *Human Relations* 7, no. 2 (1954): 117–40.

156   **result in them feeling bad:** J. P. Gerber et al., "A Social Comparison Theory Meta-analysis 60+ Years On," *Psychological Bulletin* 144, no. 2 (2018): 177–97, doi.org/10.1037/bul0000127.

156 **felt more optimistic:** M. Rodriguez, O. Ayduk, and E. Kross, "Harnessing Downward Social Comparison for Emotion Regulation" (paper in preparation).

157 **What I wished:** Suls et al., "Social Comparison."

158 **talk to them about frequency:** Peter A. McCarthy and Nexhmedin Morina, "Exploring the Association of Social Comparison with Depression and Anxiety: A Systematic Review and Meta-analysis," *Clinical Psychology and Psychotherapy* 27, no. 5 (2020), doi.org/10 .1002/cpp.2452.

158 **envelopes of cash:** Elizabeth W. Dunn et al., "Spending Money on Others Promotes Happiness," *Science* 319, no. 5870 (2008): 1687–88, doi.org/10.1126/science.1150952.

159 **question of how *much* money:** The story surrounding the relationship between income and happiness has evolved over time. An early, now classic paper suggested that happiness didn't change considerably once a certain income threshold was reached (around seventy-five thousand dollars). See Daniel Kahneman and Angus Deaton, "High Income Improves Evaluation of Life but Not Emotional Well-Being," *Proceedings of the National Academy of Sciences* 107, no. 38 (2010): 16489–93, doi .org/10.1073/pnas.1011492107. More recent research using larger data sets has upturned that finding, demonstrating that for the majority of people a linear relationship characterizes income and happiness. See Matthew A. Killingsworth, "Experienced Well-Being Rises with Income, Even Above $75,000 per Year," *Proceedings of the National Academy of Sciences* 118, no. 4 (2021), doi.org/10.1073/pnas.2016976118. Also see Matthew A. Killingsworth et al., "Income and Emotional Well-Being: A Conflict Resolved," *Proceedings of the National Academy of Sciences* 120, no. 10 (2023), doi.org/10.1073/pnas.2208661120.

159 **predictor of our emotional well-being:** Lara B. Aknin et al., "The Emotional Rewards of Prosocial Spending Are Robust and Replicable in Large Samples," *Current Directions in Psychological Science* 31, no. 6 (2022): 536–45, doi.org/10.1177/09637214221121100; Lara B. Aknin et al., "Happiness and Prosocial Behavior: An Evaluation of the Evidence," *World Happiness Report,* March 20, 2019.

160 **rich *and* poor populations:** Lara B. Aknin et al., "Prosocial Spending and Well-Being: Cross-Cultural Evidence for a Psychological Universal," *Journal of Personality and Social Psychology* 104, no. 4 (2013): 635–52, doi.org/10.1037/a0031578. For a review, see Aknin et al., "Happiness and Prosocial Behavior."

160 **Studies performed during COVID:** Mohith M. Varma and Xiaoqing Hu, "Prosocial Behaviour Reduces Unwanted Intrusions of Experimental Traumatic Memories," *Behaviour Research and Therapy* 148 (Jan. 2022): 103998, doi.org/10.1016/j.brat.2021.103998.

160   **people as selfish:** Dale T. Miller, "The Norm of Self-Interest,"
       American Psychologist 54, no. 12 (1999): 1053–60, doi.org/10.1037
       /0003-066x.54.12.1053.

160   **we have been caring for one another:** For an excellent discussion of the
       conditions under which children and adults help even when there is no
       immediate gain, see John Helliwell and Lara B. Aknin, "Expanding the
       Social Science of Happiness," *Nature Human Behaviour* 2, no. 4
       (2018): 248–52, doi.org/10.1038/s41562-018-0308-5.

160   **relinquish our own resources:** Ernst Fehr and Urs Fischbacher, "The
       Nature of Human Altruism," *Nature* 425, no. 6960 (2003): 785–91, doi
       .org/10.1038/nature02043.

160   **in our own self-interest:** Robert L. Trivers, "The Evolution of
       Reciprocal Altruism," *Quarterly Review of Biology* 46, no. 1 (1971):
       35–57; Sarah F. Brosnan and Frans B. M. de Waal, "A Proximate
       Perspective on Reciprocal Altruism," *Human Nature* 13, no. 1 (2002):
       129–52, doi.org/10.1007/s12110-002-1017-2.

161   **helps us find and keep friends:** For a discussion, see Oliver Scott Curry
       et al., "Happy to Help? A Systematic Review and Meta-analysis of the
       Effects of Performing Acts of Kindness on the Well-Being of the
       Actor," *Journal of Experimental Social Psychology* 76 (May 2018):
       320–29, doi.org/10.1016/j.jesp.2018.02.014.

161   **we feel happy:** Soyoung Q. Park et al., "A Neural Link Between
       Generosity and Happiness," *Nature Communications,* July 11, 2017,
       doi.org/10.1038/ncomms15964.

161   **Evolution gifted us:** Jessica J. Walsh et al., "Dissecting Neural
       Mechanisms of Prosocial Behaviors," *Current Opinion in
       Neurobiology* 68 (June 1, 2021): 9–14, www.sciencedirect.com
       /science/article/pii/S0959438820301744?via=ihub, doi.org/10.1016
       /j.conb.2020.11.006.

161   **practicing mindfulness and gratitude:** Curry et al., "Happy to Help?"

CHAPTER EIGHT *The Master Switch: Culture Shifters*

165   **Hollis walked into work:** Hollis is a pseudonym, used at this subject's
       request. Sources for this story were drawn from an interview with
       Hollis and the Alcoholics Anonymous website, which contains its
       "Big Book." Alcoholics Anonymous, "The Big Book | Alcoholics
       Anonymous," www.aa.org/the-big-book.

169   **our philosophy about emotions:** Jeanne L. Tsai and Magali Clobert,
       "Cultural Influences on Emotion: Established Patterns and Emerging
       Trends," in *Handbook of Cultural Psychology*, ed. Dov Cohen and
       Shinobu Kitayama, 2nd ed. (New York: The Guilford Press, 2019): 292–
       318; H. R. Markus and S. Kitayama, "Cultural Variation in the Self-

Concept," in *The Self: Interdisciplinary Approaches,* ed. Jaine Strauss and George R. Goethals (New York: Springer, 1991): 18–48.

170 **culture forms:** *APA Dictionary of Psychology* (2018), s.v. "group," dictionary.apa.org/group; Henri Tajfel et al., "Social Categorization and Intergroup Behaviour," *European Journal of Social Psychology* 1, no. 2 (1971): 149–78, doi.org/10.1002/ejsp.2420010202; Adam B. Cohen, "Many Forms of Culture," *American Psychologist* 64 (2009): 194–204, https://doi.org/10.1037/a0015308.

171 **more likely to suppress their emotions:** Foley R. A. Tsai and M. Mirazón Lahr, "The Evolution of the Diversity of Cultures," *Philosophical Transactions of the Royal Society B: Biological Sciences* 366, no. 1567 (April 12, 2011): 1080–89, www.ncbi.nlm.nih.gov/pmc /articles/PMC3049104/, doi.org/10.1098/rstb.2010.0370. Tsai and Clobert, "Cultural Influences on Emotion"; Markus and Kitayama, "Cultural Variation in the Self-Concept"; David Matsumoto et al., "Culture, Emotion Regulation, and Adjustment," *Journal of Personality and Social Psychology* 94 (2008): 925–37, doi.org /10.1037/0022-3514.94.6.925.

172 **particular context:** A. L. Kroeber and Clyde Kluckhohn, *Culture: A Critical Review of Concepts and Definition* (Cambridge, Mass.: Peabody Museum Press, 1952), https://peabody.harvard.edu/ publications/culture-critical-review-concepts-and-definitions; Hazel Rose Markus and Maryam G. Hamedani, "Sociocultural Psychology: The Dynamic Interdependence among Self Systems and Social Systems," in *Handbook of Cultural Psychology*, ed. Dov Cohen and Shinobu Kitayama, 1st ed. (New York: The Guilford Press, 2007).

172 **emotional stability:** George E. Vaillant, "Positive Emotions and the Success of Alcoholics Anonymous," *Alcoholism Treatment Quarterly* 32, no. 2-3 (June 30, 2014): 214–24, doi.org/10.1080/07347324.2014 .907032.

173 **the anthropologist Jean Briggs:** I discovered Briggs's story in this article: Michaeleen Doucleff and Jane Greenhalgh, "How Inuit Parents Teach Kids to Control Their Anger," NPR, March 13, 2019, www.npr .org/sections/goatsandsoda/2019/03/13/685533353/a-playful-way-to -teach-kids-to-control-their-anger. I then drew from Briggs's book to retell her story: Jean L. Briggs, *Never in Anger: Portrait of an Eskimo Family* (Cambridge, Mass.: Harvard University Press, 2001), 276; Jozefien De Leersnyder et al., "Cultural Regulation of Emotion: Individual, Relational, and Structural Sources," *Frontiers in Psychology* 4, no. 55 (2013), www.ncbi.nlm.nih.gov/pmc/articles/PMC3569661/, doi.org/10.3389/fpsyg.2013.00055.

175 **want others to accept us:** C. Nathan DeWall et al., "Social Exclusion and Early-Stage Interpersonal Perception: Selective Attention to Signs

of Acceptance," *Journal of Personality and Social Psychology* 96, no. 4 (2009): 729–41, doi.org/10.1037/a0014634; Solomon E. Asch, "Opinions and Social Pressure," *Scientific American*, 193 (1955).

176 **most painful emotional experiences:** Ethan Kross et al, "Social Rejection Shares Somatosensory Representations with Physical Pain," *Proceedings of the National Academy of Sciences* 108, no. 15 (2011): 6270–75, doi.org/10.1073/pnas.1102693108.

177 **Religion is "the big kahuna":** Alexandra Wormley et al., "Religion and Human Flourishing," *Journal of Positive Psychology*, Dec. 24, 2023, 1–16, doi.org/10.1080/17439760.2023.2297208; Zeve J. Marcus and Michael E. McCullough, "Does Religion Make People More Self-Controlled? A Review of Research from the Lab and Life," *Current Opinion in Psychology* 40 (Aug. 2021): 167–70, doi.org/10.1016/j.copsyc.2020.12.001; David B. Newman and Jesse Graham, "Religion and Well-Being," in *Handbook of Well-Being,* ed. Ed Diener, Shigehiro Oishi, and Louis Tay (Salt Lake City: DEF, 2018). Although the bulk of work establishing positive links between religion and well-being is based on cross-sectional correlational studies, a growing body of longitudinal research has generated findings that are consistent with the notion religion promotes well-being. According to one program of research, religion's impact on promoting emotion regulation (often referred to as "self-control" or "self-regulation" in these literatures) through its emphasis on rituals and prayer provides an explanation for how religion partly brings about these outcomes; see Marcus and McCullough, "Does Religion Make People More Self-Controlled?," for a treatment of this issue.

177 **reduce their focus on the "self":** Patty Van Cappellen et al., "Religion and Well-Being: The Mediating Role of Positive Emotions," *Journal of Happiness Studies* 17 (2016): 485–505; Allon Vishkin et al., "Religion and Spirituality Across Cultures," *Cross-Cultural Advancements in Positive Psychology* (2014): 247–69, doi.org/10.1007/978-94-017 -8950-9_13.

178 **compensatory control:** Aaron C. Kay et al., "Compensatory Control," *Current Directions in Psychological Science* 18 (2009): 264–68, doi.org/10.1111/j.1467-8721.2009.01649.x.

179 **live their values:** Center for Substance Abuse Treatment (US), "Exhibit 6-7, 12-Step Group Values and the Culture of Recovery," www.ncbi .nlm.nih.gov, 2014, www.ncbi.nlm.nih.gov/books/NBK248421/box/ch6 .box12/?report=objectonly.

181 **England's famed Football Association:** Jon Boon, "Meet Aussie Psychologist Who Helped England Banish Their Penalty Demons," *Sun,* July 4, 2018, www.thesun.co.uk/world-cup-2018/6696032/dr -pippa-grange-psychologist-england/; Russell Hope, "How England

Ended Their Penalties Curse," Sky News, July 4, 2018, news.sky.com /story/why-do-england-always-lose-at-penalties-11425022.

181   **Their goal was to create:** "FA Appoints Grange in 'Head of People' Role," ESPN.com, Nov. 10, 2017, www.espn.com/soccer/story/_/id /37538144/fa-names-pippa-grange-head-people-team-development.

182   **"Mary Poppins of football":** Jane Fryer, "The Woman Who Pulled Off That World Cup Penalty Shootout Miracle," *Mail Online,* July 4, 2018, www.dailymail.co.uk/news/article-5919187/The-woman-pulled-World -Cup-penalty-shootout-miracle.html.

182   **company's infamous founder:** Mike Isaac, "Inside Uber's Aggressive, Unrestrained Workplace Culture," *New York Times,* Feb. 22, 2017, www.nytimes.com/2017/02/22/technology/uber-workplace-culture.html.

183   **Founded in 2009, Uber:** "The History of Uber," *Investopedia,* accessed Jan. 20, 2024, www.investopedia.com/articles/personal-finance/111015 /story-uber.asp.

183   **"stand for safety":** UberValues, www.uber.com/us/en/careers/values/.

184   **successful groups:** Charles Duhigg, "What Google Learned from Its Quest to Build the Perfect Team," *New York Times,* Feb. 25, 2016, www .nytimes.com/2016/02/28/magazine/what-google-learned-from-its -quest-to-build-the-perfect-team.html.

185   **"If you know people":** Rachel Sharp, "The FA's Pippa Grange: How Culture Coaching Improves Performance," *HR,* Sept. 23, 2019, www .hrmagazine.co.uk/content/news/the-fa-s-pippa-grange-how-culture -coaching-improves-performance/.

185   **group conversations led by Pippa:** Emine Saner, "How the Psychology of the England Football Team Could Change Your Life," *Guardian,* July 10, 2018, www.theguardian.com/football/2018/jul/10/psychology -england-football-team-change-your-life-pippa-grange.

CHAPTER NINE   *From Knowing to Doing: Making Shifting Automatic*

191   **It was 3 a.m.:** I interviewed Matt on several occasions to tell his story.

194   **Gabriele Oettingen:** Angela L. Duckworth et al., "From Fantasy to Action," *Social Psychological and Personality Science* 4, no. 6 (2013): 745–53, doi.org/10.1177/1948550613476307; Gabriele Oettingen, *Rethinking Positive Thinking: Inside the New Science of Motivation* (New York: Current, 2015); Gabriele Oettingen, "Future Thought and Behaviour Change," *European Review of Social Psychology* 23, no. 1 (2012): 1–63, doi.org/10.1080/10463283.2011.643698; Character Lab, "Gabriele Oettingen—Full-Length—Educator Summit 2018," Vimeo, Nov. 14, 2018, vimeo.com/300909888; Gabriele Oettingen and Peter

Gollwitzer, "From Feeling Good to Doing Good," in *The Oxford Handbook of Positive Emotion and Psychopathology*, June Gruber, ed. (New York: Oxford University Press, 2019).

197 **"implementation intentions":** Peter M. Gollwitzer and Veronika Brandstätter, "Implementation Intentions and Effective Goal Pursuit," *Journal of Personality and Social Psychology* 73, no. 1 (1997): 186–99, doi.org/10.1037/0022-3514.73.1.186; Peter M. Gollwitzer and Gabriele Oettingen, "Implementation Intentions," in *Encyclopedia of Behavioral Medicine*, ed. Marc D. Gellman (New York: Springer, 2020), 1159–64; Peter M. Gollwitzer, "Implementation Intentions: Strong Effects of Simple Plans," *American Psychologist* 54, no. 7 (1999): 493–503, doi.org/10.1037/0003-066x.54.7.493.

198 **They work on two levels:** Maik Bieleke, Lucas Keller, and Peter M. Gollwitzer, "If-Then Planning," *European Review of Social Psychology* 32, no. 1 (2021): 88–122; Gollwitzer, "Implementation Intentions."

199 **knowledge to use in our lives:** Duckworth et al., "From Fantasy to Action"; Oettingen, *Rethinking Positive Thinking;* Oettingen, "Future Thought and Behaviour Change"; Character Lab, "Gabriele Oettingen—Full-Length—Educator Summit 2018."

201 **several studies have shown:** For a summary of select scientific articles highlighting the value of WOOP across multiple domains, see "The Science Behind WOOP," WOOP, woopmylife.org/en/science.

201 **studying and grades:** Angela L. Duckworth et al., "Self-Regulation Strategies Improve Self-Discipline in Adolescents: Benefits of Mental Contrasting and Implementation Intentions," *Educational Psychology* 31, no. 1 (2011): 17–26, doi.org/10.1080/01443410.2010.506003; Gabriele Oettingen et al., "Self-Regulation of Time Management: Mental Contrasting with Implementation Intentions," *European Journal of Social Psychology* 45, no. 2 (2015): 218–29, doi.org/10.1002/ejsp.2090l; Duckworth et al., "From Fantasy to Action."

201 **better working through of negative feelings:** Nora Rebekka Krott and Gabriele Oettingen, "Mental Contrasting of Counterfactual Fantasies Attenuates Disappointment, Regret, and Resentment," *Motivation and Emotion* 42, no. 1 (2017): 17–36, doi.org/10.1007/s11031-017-9644-4; Gabriele Oettingen et al., "Turning Fantasies About Positive and Negative Futures into Self-Improvement Goals," *Motivation and Emotion* 29, no. 4 (2005): 236–66, doi.org/10.1007/s11031-006-9016-y.

201 **healthier eating and exercise behaviors:** Gertraud Stadler et al., "Physical Activity in Women," *American Journal of Preventive Medicine* 36, no. 1 (2009): 29–34, doi.org/10.1016/j.amepre.2008.09.021; Gertraud Stadler et al., "Intervention Effects of Information and

Self-Regulation on Eating Fruits and Vegetables over Two Years," *Health Psychology* 29, no. 3 (2010): 274–83, doi.org/10.1037/a0018644.

201 **people with depression taking better care of themselves** : Anja Fritzsche et al., "Mental Contrasting with Implementation Intentions Increases Goal-Attainment in Individuals with Mild to Moderate Depression," *Cognitive Therapy and Research* 40, no. 4 (2016): 557–64.

201 **thriving relationships:** Sylviane Houssais et al., "Using Mental Contrasting with Implementation Intentions to Self-Regulate Insecurity-Based Behaviors in Relationships," *Motivation and Emotion* 37, no. 2 (2012): 224–33, doi.org/10.1007/s11031-012-9307-4.

201 **first graders:** Daniel Schunk et al., "Teaching Self-Regulation," *Nature Human Behaviour* 6 (2022): 1680–90, doi.org/10.1038/s41562-022 -01449-w.

202 *no one-size-fits-all solutions:* Fujita et al., "Smarter, Not Harder"; Bonanno and Burton, "Regulatory Flexibility."

203 **incredibly challenging time:** Baldwin Chayce et al., "Managing Emotions in Everyday Life: Why a Toolbox of Strategies Matters" (University of Michigan document).

CONCLUSION *It's 5:00 a.m., Do You Know Where Your Emotions Are?*

206 **I wrote about shooting massacres:** Kross, *Chatter,* 87–90.

ABOUT THE AUTHOR

Ethan Kross, PhD, is one of the world's leading experts on emotion regulation. An award-winning professor in the University of Michigan's top-ranked Department of Psychology and its Ross School of Business, he is the director of the Emotion & Self Control Laboratory. Kross has participated in policy discussion at the White House, spoken at TED and SXSW, and consulted with some of the world's top executives and organizations. He has been interviewed on *CBS Evening News*, *Good Morning America*, *Anderson Cooper Full Circle*, and NPR's *Morning Edition*. His pioneering research has been featured in *The New York Times*, *The New Yorker*, *The Wall Street Journal*, *USA Today*, *The New England Journal of Medicine*, and *Science*. He completed his BA at the University of Pennsylvania and his PhD at Columbia University.